Dr. GUNDRY'S
DIET EVOLUTION

TURN OFF THE GENES THAT ARE KILLING
YOU AND YOUR WAISTLINE

Steven R. Gundry,

M.D., F.A.C.S., F.A.C.C., F.A.A.P., F.C.C.P., F.A.S.A.

Director, The International Heart and Lung Institute, Palm Springs,

Founder and Director, The Center for Restorative Medicine

THREE RIVERS PRESS • NEW YORK

Three Rivers Press and the Tugboat design are registered trademarks of Random House, Inc.

Originally published in hardcover in the United States by Crown Publishers, an imprint of the Crown Publishing Group, a division of Random House, Inc., New York, in 2008.

Library of Congress Cataloging-in-Publication Data

Gundry, Steven R.
    Dr. Gundry's diet evolution: turn off the genes that are killing you and your waistline / Steven R. Gundry.—1st ed.
    1. Nutrition.   2. Health.   3. Weight loss.   4. Nutrition—Genetic aspects.
I. Title.
RA784.G862008
613.2'5—dc22   2007033201

ISBN 978-0-307-35212-5

Printed in the United States of America

Design by Nora Rosansky

10   9   8   7

First Paperback Edition

PRAISE FOR DR. GUNDRY'S DIET EVOLUTION

"A seasoned veteran of bypassing your heart blockages teaches you how to avoid his services. Dr. Gundry has crafted a wise program with a powerful track record."
  —Mehmet Oz, M.D., professor and vice chair of surgery,
  NY Presbyterian/Columbia Medical Center

"After an impressive career as a physician and surgeon in treatment of heart disease, Steve Gundry has been inspired to apply his experience, intellect, and scientific background to preventive health measures. This practical and easily readable book describes Gundry's advice for heart health and general physical improvement. While people have no choice in their heredity and genetic composition, they can work with the inherited genes, improve their personal comfort, and possibly extend life expectancy. His personal experience with control of obesity contains timely advice on this issue, which is affecting modern society."
  —Denton A. Cooley, M.D., president and surgeon-in-chief,
  Texas Heart Institute, Houston, Texas

"After my bypass surgery, I read every diet and health book I could find. Dr. Gundry's book is revolutionary because its new science is presented in a creative, fun, and easy-to-understand way. You'll want to take immediate action for long-term results; the plan is simple and life-changing."
  —Greg Renker, cofounder of Guthy-Renker

"Dr. Gundry's thirty years of experience in the field of medicine have made him an invaluable resource to people all over the world. His provocative strategies will help you achieve the sustainable, healthy, and vital life you deserve."
  —Anthony Robbins, bestselling author of
  Awaken the Giant Within and *Unlimited Power*

To my wife, Penny, my soul mate
and the best person I have ever known.

To my mother, Bev, who taught me
to think and cook, and to my father, Bob,
who taught me to love people. They are
the best parents I have ever met.

DISCLAIMER

The information presented in this work is in no way intended as medical advice or as a substitute for medical counseling. The information should be used in conjunction with the guidance and care of your physician. Consult your physician before beginning this program as you would any weight-loss or weight-maintenance program. Your physician should be aware of all medical conditions that you may have, as well as the medications and supplements you are taking. As with any weight-loss plan, the information here should not be used by patients on dialysis or by pregnant or nursing mothers.

JUST A MINUTE . . .

Before you begin *Diet Evolution*, I'd like you to meet some of the people who've tried the program. Flip to page 283 to find out how they're doing.

Burt Kaplan

At 76, Burt Kaplan was getting old—or so he thought. Despite eating heart-healthy cereal with a banana and low-fat meals, Burt's weight had skyrocketed out of control. He was taking three medications to control his blood pressure and angina, and until he walked into my office, he had never been told he was an insulin-resistant Type 2 diabetic . . .

Judith Rhode

Judith Rhode needed emergency surgery following a heart attack that left her with a leaky heart valve and a severely weakened heart. Paradoxically, the heart attack was one of the best things that could have occurred. Before it happened, she suffered from severe insulin-dependent diabetes, obesity, high blood pressure, high cholesterol, and leg and hip pain that often caused her to use a walker. After her operation, she started Diet Evolution . . .

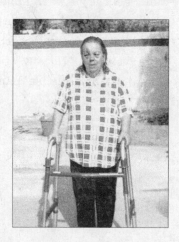

Sandra Hall

Sandra Hall is a 30-ish mother who had failed at every diet she had ever tried. She suffered from high cholesterol, but when her doctor told her she would have to start taking high blood pressure medicines at her young age, she'd had enough. When she saw how well a friend was doing with Diet Evolution, she asked her to make a copy of the small handout I'd given that friend, which briefly explained the diet. So without ever meeting or talking to me, Sandra embarked on the program . . .

Margo Hamilton

Two hip replacements in your late 40s and early 50s as a result of obesity ought to get your attention. How about undiagnosed high blood pressure, high cholesterol, and insulin-resistant diabetes? Despite eating "healthy," Margo's weight kept piling on; moreover, every exercise program was unbearable. Margo didn't believe me when I told her she wouldn't be hungry, would have lots of energy, and would experience zero cravings if she tried Diet Evolution. "No way," she said! And lose weight without exercise? Impossible . . .

CONTENTS

Dr. GUNDRY'S
DIET EVOLUTION

INTRODUCTION

Is this you? Your skin is clear and unwrinkled, your slim body moves with ease and grace, and you're blessed with strength, stamina, and good health. You look 45 tops, but here's the zinger: you're really 70. No way, you say.

Forget 70; let me introduce you to Michelle (a pseudonym, as are most patient names). At our first meeting, I saw a striking, thin, erect woman who appeared to be about 65. I looked at the chart again and thought I had the wrong examination room. Her age was listed as 95! Michelle had recently seen me on television and said I was the first doctor who talked like one she met when she was 20, who had changed her life. His message was succinct: Go home and throw out every white food in your pantry and never eat such food again. For 75 years, she had done just that. In that time, she had buried two husbands, including a physician, who told her that her eating habits were crazy. He was long dead, but here she was, her perfect skin radiating health. Unlike many women in Palm Springs, she had never had plastic surgery.

When I took Michelle's temperature, it was 95 degrees, meaning she had a low, and therefore efficient, metabolic rate. Her blood pressure was an exemplary 95/55. When her blood samples were analyzed, as I expected, they were perfect: low cholesterol, no evidence of inflammation. An active businesswoman, Michelle is now 96 and still shows no signs of aging. Why would she? Her longevity genes are activated, preserving her from harm. Rather than being in survival mode, she exists in what I call "perfect efficiency," which is what I hope to attain—and the same goal I have for you. Live long and prosper, Michelle.

Is this you more like you? You weigh more than you did in high school. Perhaps a lot more. You pop one or more daily pills for high blood pressure, high cholesterol, diabetes, acid reflux, depression, and/or arthritis. You may be sending your dermatologist's children through college by having benign "skin tags" burned off your neck and armpits. While you're there, you may get treated for adult acne. If you're a woman, your hair may be thinning. Perhaps you've had some colon polyps and/or breast lumps removed. Am I hitting a little too close to home? If so, what if I suggested that even if you do not already have diabetes, hypertension, heart and vascular disease, cancer, or another life-threatening disease, you've got a lot in common with people who do?

What if I could demonstrate that you and they have all unwittingly set into motion what I call "killer genes," which caused these and other unforeseen consequences? Preposterous? Until six years ago, I certainly would have thought so, but now I am convinced that these seemingly chance events are, for a great many of us, predictable outcomes. And, to a large extent, it's our Western diet and lifestyle that are making us sick and ultimately killing us—although paradoxically, as you'll soon learn, they suit our genes just fine.

Sadly, as a heart surgeon, I don't see many people like Michelle—at least initially. Many of my patients are severely ill and often prematurely aged. Most are also overweight. Why some people like Michelle seem to have sipped from the fountain of youth even as they near the century mark, while most of us are wrinkled, arthritic, overweight, and plagued with complaints by the time we're eligible for AARP membership, has long been a mystery. Is it luck? Good genes? Over the last five years I have unraveled much of that mystery. Yes, our genes play a major role, but not in the way we have been led to believe. Michelle did not inherit "good genes." Quite the opposite: since her life-changing encounter 75 years ago, she has been instructing her genes to "be good"! I've also discovered how to do the same no matter what genetic cards a person has been dealt. What I have found argues powerfully for a major and immediate change in lifestyle. *Dr. Gundry's Diet Evolution* is based on radically simple, but simply radical changes that will make you slimmer, fitter, and healthier. You may well become another Michelle!

By now you're probably thinking I'm a fanatic trying to spread the latest theory of doom and gloom. Or that I'm the equivalent of a modern-day snake-oil salesman determined to dupe you into blowing your hard-earned dollars on worthless "natural cures and supplements." I can assure you that I am neither. In fact, my credentials would suggest that I am the most unlikely person to suggest that killer genes are responsible for excess weight and a host of other ills.

I'm in the human survival business. Until six years ago, I primarily flexed my survival muscles as a heart surgeon and researcher on how to keep heart cells alive under stress, operating on infants with malformed hearts and adults with blocked coronary arteries or valves. I've been peering inside bodies and especially the heart's blood vessels for the past 30 years as the former Professor of Surgery and Pediatrics in Cardiothoracic Surgery and Head of Cardiothoracic Surgery at Loma Linda University School of Medicine.

I'm also the inventor of the Gundry Retrograde Cardioplegia Cannula, one of the most widely used devices to keep the heart muscle alive during open-heart surgery by delivering heart-protective ingredients "backward" through the veins of the heart. Before, everyone else was trying to push them forward past arterial blockages. My "backward" idea was initially called crazy—"like giving the heart an enema!" said prominent heart surgeon Dr. Denton Cooley. But today, my device is the gold standard for protecting the heart during surgery. Why? Because I refused to accept conventional wisdom. Millions of patients later, conventional wisdom has been overturned.

I also hold patents on numerous other devices to repair leaky heart valves or "sew" new blood vessels into the heart without sutures. Together, my colleague Leonard Bailey and I have performed more infant and pediatric heart transplants than anyone in the world. My laboratory holds the record for producing the longest surviving pig-to-baboon heart transplant—twenty-eight days—when others claimed it could last only hours. I was one of the original twenty surgeons who tested the first successful artificial heart, one of the first surgeons to use robots in operations, and the first to design and perform heart-valve operations through two-inch holes!

You could say I'm a maverick. I've always looked at problems of the heart, and survival, from a different perspective. Perhaps that's why I've had the honor of being inducted as a Fellow of the American Surgical Association, comprising the top 500 surgeons in the world, and have consistently been selected by a poll of 10,000 of my peers as one of Castle Connolly's top American doctors. I've also published more than 300 articles, abstracts, and book chapters on my research and techniques.

Another thing about myself, but of this I am anything but proud. I was obese. Yes, that's right, an obese heart surgeon. I was also hypertensive and insulin resistant, and I suffered from migraine headaches and arthritis. Even though I ran twenty miles a week, visited the gym daily, ate "healthy," and always had a Diet Coke in my hand, I was still obese. I lost weight on every diet that came down the pike, and then rationalized its inevitable return. Until I discovered how to communicate with my genes, I couldn't seem to deal with the extra weight, high cholesterol, arthritis, and tendency to diabetes—the very same health issues many of you are facing.

One more claim to fame: I was the medical consultant and also had a bit part in the movie *The Doctor*, starring William Hurt, about a heart surgeon who changes his ways when he develops a potentially life-threatening disease. As it turns out, the premise of the film couldn't have been more prophetic.

Surgeons are trained to fix problems, usually with an operation. Six years ago, when I was at Loma Linda, a guy I'll call "Big Ed" strode into my office, and a day later, my career and my life made a sharp U-turn. He was carrying an angiogram—a movie of the heart's arteries that, among other things, reveals the presence of cholesterol plaques or blockages. Big Ed, who was then in his late 40s, could serve as the poster child for Harley-Davidson bikers: you know, big beer gut, ponytail, and two days' growth of beard. He had been told that his condition was inoperable and his cardiologist referred him to me as someone who takes cases no one else wants. But, after reviewing Big Ed's angiogram, sadly I had to agree with the other surgeons: inoperable.

Crushed but determined, Big Ed then informed me that since that angiogram had been done six months earlier, he'd lost 45 pounds and had been taking "handfuls of herbs and supplements" several times a day. Maybe, he suggested, his efforts had cleaned up his arteries. His story was interesting, but Big Ed had only downsized to 260 pounds—like the Hummer 2 compared to the original.

As I stroked my professorial beard and smiled knowingly, I told him his worthless supplements only made "expensive urine," as I was fond of saying then. While his weight loss was commendable, it wasn't going to improve the state of his blood vessels. But Big Ed made me an offer: since neither of us had anything to lose, how about we repeat his angiogram to see if anything had changed?

Well, Big Ed got his quintuple bypass later that week. Although his arteries were still very diseased, remarkably most of the blockages had shrunk by more than 50 percent, meaning that there were now places to "plug in" new blood vessels. In all my years, I had never seen such a reduction in blockages. As a surgeon I was delighted, but the researcher in me was even more intrigued. So after Big Ed left the hospital, I asked him to bring in his treasure trove of vitamins, minerals, and herbs. As I listened to how he had constructed his "diet," I was transported back thirty years to when I researched this field for my honors thesis on Human Evolutionary Biology at Yale. After medical school, I applied this knowledge as a Clinical Associate at The National Institutes of Health, where I developed methods and treatments to prevent and reverse damage in coronary arteries and heart muscle cells.

Our encounter started a personal quest that up-ended most of my preconceived notions about the underlying causes of heart disease, diabetes, cancer, arthritis, obesity . . . and you name it. I dug up my college thesis to try my own diet based on principles of human evolution. Using sophisticated bioassays of my blood from a renowned laboratory, I experimented with foods, supplements, and exercise. The results were amazing. I lost 50 pounds over the first year; I've since lost another 25, but more important, I restored my body's normal cellular functioning. As just one example, my LDL ("bad" cholesterol) dropped more than 100 points while my HDL ("good" cholesterol) increased 150 percent. All without drugs. My blood pressure, which was once 145/95, is now 90/50. As my program progressed, the numerous so-called benign skin tags I had had for years fell off. My migraine headaches vanished. I could go on and on about the benefits.

What happened next is still a blur. I resigned my position at Loma Linda University and moved to Palm Springs to establish The International Heart and Lung Institute, and within it, The Center for Restorative Medicine. I began to offer my Diet Evolution program to patients I had operated on, in the hope that it would enable them to avoid a second operation. My office nurse, nurse practitioner, and office manager all started the program and lost 72 pounds, 45 pounds, and 42 pounds, respectively.

Then one fateful day, I was referred a patient with 90 percent blockage in the main artery to the brain. He needed a carotid endarterectomy, which entails opening up the artery, removing the plaque, and sewing it back up. His case was especially tricky because the lesion was relatively inaccessible, requiring the use of an experimental technique, so I suggested he try Diet Evolution while we consulted with the surgeon who pioneered the technique and ran some more tests.

Two months later, he was down 12 pounds, and while doing his pre-op examination, I found I couldn't hear the characteristic noise this blockage makes in the neck artery. When we did a new scan, amazingly the blockage was reduced to 30 percent. The successful results made it clear that Big Ed's outcome was not a fluke. (You can read this "success story" on page 89 in "Cancel the Vascular Surgery," and throughout the book you'll meet many of my other patients who have reversed their health problems and shed excess weight.) As each new patient's eating habits, lifestyle, and blood work were analyzed, an intriguing pattern emerged: everyone—and I do mean everyone—responded in exactly the same manner to certain foods and behaviors.

Soon, other people I have come to call Club members began arriving, asking me to teach them about Diet Evolution. Although I never advertised this program, word of mouth has changed my practice forever. Three days a week I teach these members how to stay away from me, earning the nickname "No More Mr. Knife-Guy." The remaining two days I operate on people who don't listen to me or haven't yet heard my message. Most of my surgery patients soon become Club members, too. With remarkable consistency, not only do their cholesterol problems disappear but also so do their weight problems, hypertension, diabetes, and arthritis. We've even established preventive and treatment programs at a local cancer center.

Most of what I've learned from the research on myself and thousands of volunteers flies in the face of what health professionals have been taught about health, diet, and chronic disease. It confirms, though, that we are genetically programmed to eat a diet that helps quickly reproduce our genes but is actually killing us—the body that houses them. This deadly diet activates a host of killer genes designed to ensure that you don't stick around very long, and is also in large part the cause of the epidemic of obesity sweeping the nation and spreading viruslike to the rest of the world.

I am offering you the opportunity to take a different path. When you've finished Part One of this book, you'll have a basic understanding of how your genes work and how you can reprogram them. You'll do this by slowly *re-evolving* your diet to take advantage of how evolution has programmed your genes to respond to the things you do and eat.

By sending the *right* messages to your genes, you can restore your health, lose excess weight, and likely extend your life. In Part One, you'll learn how to:

- Identify the foods and behaviors that leave you open to disease and chronic health conditions, and how to eliminate or moderate them.
- Reintroduce the key foods, behaviors, and nutritional compounds that promote health and longevity.
- Turn off those killer genes.

Because 90 percent of the cells in your body are replaced every three months, the great news is that you can create a "new" you in as little as 90 days. In Part Two, I'll provide you with an easy to follow three-phase program that lays the groundwork for achieving life-changing results in as little as three months. Welcome to The Club!

Part One

A NEW

WAY TO

COMMUNICATE

WITH

YOUR GENES

Chapter 1

YOUR GENES ARE RUNNING THE SHOW

If you're anything like me, I know you're champing at the bit to get going on Diet Evolution, but hold your horses. I've found that most of us can stick to a program only if we understand how and why we got to our present state of affairs. The next four chapters will do just that.

You can thank Mom and Dad for your beautiful baby blues, as well as your hair color, height, and build. All these traits were encoded in copies of their genes—half of them her's, the other half his—that now reside in your body. Any children you have will in turn have copies of half of your genes and half of your partner's, and so on through generations to come. Determining our appearance and myriad other characteristics is just one way in which our genes rule our lives. As you'll soon learn, they also play a more clandestine role.

The answer to why almost all attempts to lose weight fail resides in your genes and the lies they've been feeding you. Once you understand how genes behave, and how their behavior controls your behavior, you'll be rewarded with an understanding of how you can lose those extra pounds, regain your health and vitality, and set the stage for a long, vigorous life. I'm sure you have heard that if you feed your genes right, all your health problems will disappear. But here you are stuck in the same old rut. I'm going to prove to you that you're in poor health and overweight because you're feeding your genes exactly what they want. Let me assure you, your genes are using you for their health and longevity, not yours. In fact, your longevity stands in their way. So sit back, open your mind, and be willing to suspend everything you've ever known about yourself. But first let's get down to basics.

GENES AS MICROCOMPUTERS

I've found that it helps to think of genes as tiny computer programs. On your computer, the letter *A*, for example, is "coded" as 1000001. Whenever I strike *A* on the keyboard, this code of 0's and 1's tells my computer to make an *A* appear on the screen. Think of your body as a computer and your genes as its operating system. Genes are pieces of information composed of hundreds or even thousands of sugars and proteins arranged in sequences that "spell out" the process the gene wants a cell to perform. They contain all the information needed to tell each cell what to do, but in this case, the keys that turn genes on or off are circulating hormones, neurotransmitters, and numerous other information carriers, particularly compounds in food. To continue the computer metaphor, these codes work much as typing an Internet address into your Web browser "tells" it what information you want.

Your genes are totally dependent on your body to protect them and house them until a new computer, in the form of your offspring, can take on these tasks. We need our genes as much as they need us: your body only works when its operating system—your genes—tells it what to do. Over millions of years, the communication between our genes and the human body has undergone continuous improvements and upgrades, all to ensure perpetuation of the genes. Despite myriad improvements—much like Vista replacing Windows, or Panther replacing Tiger—the basic operating system of your genetic program remains the same.

Okay, you head-shakers, if all this just seems too fantastic to believe, let me ask you why you have no trouble accepting the fact that you can watch *Oprah* on your cellphone, thanks to invisible electrical impulses from outer space that communicate with cell towers to activate tiny computer programs in your handheld appliance. An invisible but equally powerful, comparable process also happens in our bodies. Why do you think that after two months female dorm mates find their menstrual periods have synchronized? Their hormones sent messages through thin air—rather like text messaging—for them to ovulate at the same time.[1] Hormones or plant chemicals stimulate genes to either switch on or off, determining what's going to happen in each cell, each part of the body, and each person.

YOUR GENETIC AUTOPILOT

Indeed, whether you realize it or not, your fate is being controlled by a hidden system that runs most of your cellular, hormonal, nervous system, and aging processes without your conscious input. It does this, so that the "thinking you" doesn't screw up the process of moving genetic material forward in time. If you've seen Stanley Kubrick's classic *2001: A Space Odyssey*, in which intergalactic astronauts travel in a spaceship autopiloted by a computer they nickname "Hal," you can envision what's going on in your body. Hal takes care of running all the spacecraft's functions with minimal human input, but when the astronauts try to take over the vehicle, they realize that Hal is in total control and they're just along for the ride. When they attempt to thwart Hal, he tries to destroy them! Your autopilot is usually invisible, but I'm convinced that we constantly receive warning messages from it when processes head in the wrong direction.

For example, right now you're breathing without thinking about it, but the moment you do, you become conscious of your inhalations and exhalations. Now, let's do a quick test. Stop reading and time yourself as you hold your breath as long as you can. Were you able to go for two minutes? Up to three minutes? Or were you gasping for air at thirty seconds? You mean you were forced to take a breath? How can that be? Aren't you in charge of your body? I ask you to perform a simple task such as holding your breath, and now you're telling me that it was not under your control. Congratulations, you've just met your autopilot, your second self. This mechanism is there for a very important reason: to keep your genetic code from being wasted, in case you get the crazy notion to stop breathing when you're asleep or unconscious, among other things. That's just one of many examples of how Hal's mission is more important than yours!

We now realize that this operational program is far more than a second nervous system over which we have little or no control. Rather, it's a highly sophisticated system of specialized cells and the genes within them, including those that produce and sense hormones, govern your immune system, line your gut—all in constant communication with each other without your conscious input. If you're still having trouble with this concept, consider the autopilot program on an airplane. Once the human pilot inputs data about the destination into a computer, the autopilot controls direction, speed, thrust, flaps, pitch, and yaw. The autopilot doesn't "see" where the plane is going, doesn't "feel" how fast it is flying, but based on information sent from sensors in the plane or bouncing off satellites, it pretty much "knows" where it is in space and time, allowing it to "fly" the plane and land safely. But input the wrong information and the autopilot will dutifully fly you directly into a mountain, because that's where you told it to go. You'll understand more what I mean by this in the next chapter, when I'll explain how your current way of eating is giving your autopilot the wrong information, programming you for disaster.

PROGRAMMED FOR SURVIVAL

At the most basic level, our genes program us to do three things, all with the same purpose: to reproduce. That's right, your genetic code wants nothing more than to keep living. Notice that I didn't say it wants to keep you alive. Your genes are only using you as a host. When they're done with you, they throw the switch, activating the self-destruct "killer genes," and your days are numbered. Bummer! Now let's look at the three directives for which we are genetically hardwired.

Find and Conserve Energy This one sounds like a no-brainer, but for almost all animals (and plants, too), obtaining adequate nutrition has traditionally been an iffy proposition. Starvation not only kills directly, but also weakens an animal, leaving it more vulnerable to a predator. If you're dead, you can't pass on your genes to the next generation. So obtaining adequate calories with the least expenditure of effort is a primary driving force for all living things. If an animal is consistently in overdrive, burning up mucho calories in pursuit of food, its genes get the message that this combination of genetic material isn't a very good design and therefore should be weeded out. What this means for you is that your genes are programmed to send messages to your body to keep your metabolism low and to eat high-calorie foods.

Our genes are not only willing participants in this almost irresistible force to eat high-calorie foods, they also drive you to this behavior. Just as you can't suppress the eventual need to take a breath, the drive to conserve energy and obtain the most energy in the form of calories for the least effort is overpowering. Just think about that the next time you get in the drive-through line at a fast-food joint. Then consider this: Studies of overweight people show that if hunger strikes when there is no food in the house, they often won't make the effort to go out and get it.[2]

Your computer program also initiates a self-destruct process when it senses that you're overconsuming fuel that will be needed by the next generation to continue the genetic mission. Remember, in any ecosystem, there is only so much food; if you appear to be taking more than your fair share, you're weeded out as a threat!

Avoid Injury and Pain With the possible exception of teenagers wanting to get their bodies pierced in places you didn't know could be pierced, animals have an innate desire to avoid pain; that's because in most circumstances, pain equates with the potential for injury. If an animal is wounded, it becomes predator bait and its genes die with it. Our hard-wired impulse to avoid pain helps protect us from injury, without the conscious brain having to waste valuable split seconds "thinking" about a response. So associated is pain with injury, and so severe the consequences of injury on the chances of survival, that we avoid pain at almost all costs.

Find Pleasure All animals are programmed to seek out pleasurable experiences, but copulation itself is not the only source of pleasure. The entire mating ritual stimulates feel-good receptors in the brain, which give rise to feelings of pleasure.[3] Nor is it a coincidence that these receptors live right next door to those for sex hormones. Behaviors or compounds that stimulate these pleasure centers always override the other directives. Stimulate an electrode placed in any animal's brain pleasure center and it won't eat, sleep, or even avoid pain. Remarkably, the animal will even continue to press a button for this stimulus, avoiding food or water, until it blissfully starves to death.[4] Likewise, given a choice between a pellet of food and a pellet of cocaine, a rat opts for cocaine until it dies.[5] Now that's hardwiring!

Another more commonly used—and abused—pleasure-stimulating plant compound is found in fruit: sugar. It's in the plant's survival interest to have animals eat its fruit, including the seeds. But how does a plant make an animal eat the fruit instead of its leaves? Easy; just stimulate the animal's pleasure center. Rats don't overeat rat chow, yet give them sugar or milk chocolate and wham, they can't stop gobbling.[6] Most of us are no different: we keep eating sugar-laden food until we're dead. And here's the worst part—sugar is exactly what your genes want you to eat.

And here's why. You have a certain type of shoulder joint, shared only with great apes, that enables apes to hang from and move under branches, including the flimsy ends of branches where most fruit hangs out of reach to other primates. Every study comparing apes to other primates confirms that we are designed to fatten up on fruit others can't reach—on a seasonal basis.[7,8] This genetic and anatomic design foreshadows everything that has recently gone wrong with our current lifestyle, so stay with me to learn about its import in the coming chapters.

YOUR GENES VS. THE REST OF YOU

As you're coming to understand, like it or not, your genes have constructed you to achieve three main goals:

1. Deliver genes into the future by reproducing.
2. Ensure the survival of your genetic copies or other similar copies.
3. After accomplishing #1 and #2, get out of the way so you don't compete for limited resources with your offspring.

Now here is the clincher. The faster you accomplish these goals, the better things are from your genes' point of view.

To accomplish #1 and #2, you're programmed to find the most food using the least expenditure of energy by seeking out three pleasurable tastes: sugar, fat, and salt. From your genes' viewpoint, if food is nutritious (by this I mean high-calorie, not necessarily good for you) and enjoyable, you'll eat more of it, enhancing *their* chances for survival. As an example of how intricate this energy-conserving design is, as you get older, your adrenal glands produce progressively less and less DHEA, a hormone responsible for stimulating muscle-mass production. Under orders from your genes, your DHEA level drops as you age, causing you to lose about one-third of your muscle mass between the ages of 30 and 60. While distressing to us, the result is that you don't have to eat as much—muscles require calories—and violate rule #3: Don't compete for resources with your offspring.

WHEN PICKINGS ARE SLIM

Whenever conditions are unfavorable for the survival of offspring, a form of genetic override kicks in. During periods of food scarcity, pregnant women miscarry male fetuses in much higher numbers than female ones.[9,10] This occurs in part because you need only one man to fertilize lots of females' eggs, but many females to populate the society. Also, because men generally are bigger and eat more, smaller, less muscular females put less strain on the food supply in difficult times. Scary to imagine that your genetic computer knows the sex of your unborn child and can instruct your body to jettison your precious cargo if it doesn't fit "the plan."

Not convinced yet? Women who don't have enough fat stores to successfully carry a pregnancy to term without eating don't ovulate.[11] Why waste an egg if those genes won't be delivered into the future? That's why very thin female athletes and many anorexic women can't get pregnant. What happens when a mother is undernourished during her pregnancy? Her infant will have hunger genes that are permanently upregulated, meaning they are turned on. And here's the amazing thing: the hunger genes of those children's children and grandchildren are similarly activated for life.[12] Often called "thrifty genes," these genes anticipate another period of scarcity and want to ensure that when it comes, the hungriest offspring with the largest fat stores will survive.

GENES FOR HIRE

Neglecting or flaunting these three principles, particularly not competing with your offspring, activates a second tier of genetic programming, made up of "killer genes." Let me assure you, this is not a theory. Scientists have known about the behavior (if not the actual existence) of these genes for years. Killer genes clearly deserve their label for the chillingly—yet miraculously—methodical and predictable way they activate to carry out their instructions. I suspect that soon you, too, will be awed by their power.

Recent discoveries have upended the mechanisms of aging and death that most of us have been taught as immutable. First, the usual "wear and tear" theory of aging is (pardon the expression) dead wrong. Until recently, researchers were convinced that as body parts and cells wear out and simultaneously built-in repair systems become less effective, humans and most other animals slowly but inevitably age. Cancer cells, which have been lurking like looters in the background, are now free to run amok. Like your car's brake pads, your joints wear down. Crud builds up in your arteries, and eventually the major arteries to your heart or brain shut down. A tidy theory, but it's incorrect. Aging and death are *not* preordained, but like everything else, they are essential components of our genetic programming that is still evolving.

Here's the shocker: Creatures such as sharks and alligators that date back hundreds of millions of years have no aging mechanism and appear to have *no finite age limit.* Another example: Cell lines of cancer tumors have survived unchanged in petri dishes for more than sixty years, long after their human hosts have died.[13] All of this suggests that, rather than being preordained, aging and death are "ordered" to occur, and in most species ought to occur, to make it easier for plants and animals bearing new copies of genes to have a shot at continuing the lineage and to encourage a multiplicity of new genes capable of adapting to new environmental challenges.

Killer genes exist to get rid of individuals who have outlived their usefulness or, just as important, threaten the future of the system. Remember Hal, the autopilot in charge of the spaceship in *2001*? Think how a well-designed autopilot would view the input it receives from the spaceship (you) about how the trip is going. If your activity decreases, Hal would assume that you're injured and immediately reduce the amount of food your muscles receive by making your muscles less able to take up glucose, a condition known as insulin resistance. (As you will learn in Chapter 2, insulin carries glucose, or blood sugar, to the cells, where your body can use it for energy. Excess glucose is converted to fat and stored for future use.)

A case in point: Athletes prescribed bed rest for 48 hours become insulin resistant.[14] Why would this happen? To reduce the utilization of fuel by the main consumers, the muscles. Wow! Two days of inactivity make you insulin resistant, so you'll conserve fuel. Think of the unwitting messages we send our genes by simple behavior changes. What if you're less active, but continue to consume a lot of calories? Well, if you're a bear about to go into hibernation, fine. They gorge on an abundance of berries and salmon, taking advantage of insulin-resistance activation to fatten up and simultaneously store water. They live off this stored fuel and water for the next five months or so. When you nosh on foods high in sugar and starches, you, too, activate insulin resistance, but continue and you become overweight, which your genes perceive as threatening the food supply of others. Likewise, as your muscle mass decreases with age, if you continue to eat like a teenager but are inactive, your genes assume that you're taking more than your share without contributing to defense or food collection. In either case, your behavior unwittingly activates killer genes.

A "RUSTY" BODY

Wild animals low in the pecking order have to work harder to get food and usually settle for the worst parts of the kill or the least nutritive leaves. Both factors increase the level of oxidative stress on the system, just as smoking does to humans. (In simple terms, oxidation is caused by free radicals that "rust" your body, much as metal degrades when exposed to oxygen.) Ultraathletes also exhibit high levels of oxidative stress.[15] If your autopilot detects heightened oxidative stress, it perceives you as an unsuccessful animal, harboring genes that are not worth passing on. Killer genes are being activated in smokers and marathoners as well as folks who live on foods full of refined grains, sugar, and trans fats and deficient in micronutrients. Most patients hospitalized with heart disease or diabetes show a similar nutritional profile: they're overfed but undernourished, with remarkably low levels of quality protein and vitamins and minerals revealed in their blood work. They wound up in the hospital because their genes are trying to dispose of them.

Okay, we've identified several areas of behavior (overeating, inactivity, overexercising, and smoking) that might trigger killer genes to go on the warpath, but how would they go about activating the destruction of the body they inhabit?

WHEN GOOD TURNS BAD

Now listen up, because the first time you read the next sentence, you'll think I have made a mistake. But what I'm about to tell you is the sad truth. Our contemporary foods are so "good" for you, that they actually become "bad" for you. That's because, after a set period of time, it is necessary to get you out of the way. If you keep eating these foods after you've done your job of creating and raising a new generation, then, and only then, these foods activate killer genes to keep the overall genetic program moving forward.

The name for this phenomenon is *genetic pleiotrophy*, meaning genes that activate one sequence of events during part of the life cycle activate the opposite events when called upon to change direction. It's a brilliant plan, administered by your computer program. The foods of developed nations are so "good" for your genes that they are "bad" for you and our entire society. Your genes encourage you to eat certain foods, all for a nefarious purpose: their survival, not yours. But here's the important news: you *can* outwit them. Remember how Hansel and Gretel tricked the wicked witch, who fortunately was almost blind? When she put her hand through the bars of their cage to feel if they were plump enough to eat, they substituted chicken bones for their fingers. In *Diet Evolution*, you'll learn how to:

1. Fool your genes into thinking that you're not fat enough to kill yet.
2. Convince them you're not working overtime, struggling to survive.
3. Get them to reverse their effect so they undo the damage they—with your help—have already done to your body.

Make sure you understand this concept, which contradicts other genetic-code diet books. Feeding your genes what *they* want is how you remained trapped in your overweight or poorly functioning body. Feeding them what is good for *you* and eliminating what is bad for you will get them on your team. In the next chapters, I'm going to teach you how to align your genes' purposes with your own.

THE SEDUCTIVE WESTERN DIET

When a primitive culture adopts the Western diet, and particularly refined carbohydrates, within one generation its people begin to experience the typical diseases of the civilized world.[16] Hypertension, diabetes, heart disease, arthritis, cancer, and colitis—diseases currently unknown or rare in such cultures—become rampant. So consistent is this phenomenon that it is referred to as "the rule of twenty years."[17] Think about this: research has shown that skulls of primitive man, and even fossils of our early forbears, always had perfect dental arches and straight teeth; but when primitive cultures adopt the Western diet, within just one generation orthodontic problems appear along with the usual cast of chronic diseases.[18]

You've been told repeatedly that our usual way of eating is terrible for you, and is the cause of all your health and weight problems, yet you're hooked on it, right? You're in good company. When Pacific Islanders, Native Americans, and Aleuts are exposed to the Western diet, they convert immediately. So do baboons, bears in Yellowstone Park, our dogs and cats, rats and monkeys, and other animals. By now, you understand that it's your genes, your programming, that make you want this stuff. But if it's so bad for you, why would your genes want you to eat it?

Let me say it one more time: because it's good for *them*! Your body grows faster and becomes stronger; females produce babies at a younger age and have more of them. But all that high-calorie food has a price, because of the "calorie counter" for each species on earth. Within our master computer is a program that monitors the number of calories we consume and compares it to a standard that allows each human (like all other animals studied) to grow, reproduce and raise children, and then get out of the way in order to conserve precious food resources. After exposure to these modern foods, you should be able to accomplish just about everything your genes need done, meaning replacing yourself, before you have to begin your exit in about twenty years. As just one example, the Pima Indians of the American Southwest were once long-

Bye-Bye Wheelchair and Gastric Bypass

. . .

Our local bariatric surgeon, Dr. Bobby, referred a woman in her early 30s to me for help. Rachel had become completely disabled by her weight and had to get around on a motorized wheelchair. She weighed 340 pounds, making it impossible for Dr. Bobby to perform gastric bypass safely. Rachel had every medical problem in the book: high blood pressure, diabetes, severe arthritis, and abnormal heart rhythms. It was difficult for her to understand that the budget-conscious food she was eating—beans, rice, macaroni and cheese, and pretzels—was killing her. But she was motivated by Dr. Bobby's promise that that if she dropped under 300 pounds, he would perform her stomach stapling surgery.

When Rachel hit 299 pounds, I called Dr. Bobby and told him that she was ready for surgery. About a month later, after losing another 10 pounds, Rachel walked into my office supported by a cane. She asked me if she had to go through with the gastric bypass surgery. I told her it was her decision, but that anyone who was doing what she had done probably didn't need it.

Now at 275 pounds, Rachel continues to follow Diet Evolution and is losing 4 pounds every month, as she heads toward a new life. I found out recently that her ex-husband had visited and asked her to remarry him and move back to Indianapolis, her hometown. She's leaving the motorized wheelchair behind.

lived, but once they switched to the Western diet, their birth rate doubled, but adults now die extremely prematurely from diabetes and heart disease.[18,19]

THE *REAL* CAUSE OF VASCULAR DISEASE

Now that you understand this concept, imagine you were the master genetic computer programmer. To activate killer genes, what programs would you write? Let's start with coronary artery disease, and vascular disease in general, which results from "calluses" accumulating in our blood vessels. Traditional thinking holds that the Western high-fat diet causes fat to stick to our arteries. Maybe so, but why doesn't this fat stick to all of our arteries? Why just the ones in which blockage is lethal or at least incapacitating?

Why are these distinctions important? We physicians currently like the public to think that if your coronary arteries have crud in them, there's a strong likelihood that arteries in your brain, legs, and every other part of your body do as well. Logical, yes; accurate, no. Along with most of my colleagues, I was taught this simplistic approach, but if it is so right, why have you never heard of someone having a "nose attack"? That's right, if this crud is building up all over the place, in every artery, why doesn't a patient's nose drop off?

Ridiculous? Of course, but it makes perfect sense if the crud-accumulating-throughout-the-body explanation of heart disease is correct. But it's not correct in the vast majority of cases. Arterial blockages occur only in areas specifically designated as appropriate targets for your killer genes. Close off coronary arteries—perfect! A heart attack is a great way to get rid of you or at least slow you down. Clog up an artery to your brain—wham, a stroke! Another great design to end your reign. How about your leg arteries: sure, you can't walk or run because of the pain in your legs. All are perfect ways to cut you off from your food supply and make you a target for predators.

THE *REAL* CAUSE OF *ALL* OUR MODERN DISEASES?

Did you know that hypertension simply doesn't exist in societies that don't eat the Western diet or lots of refined grains? But in our culture, after we eat such foods for about twenty years, hypertension rears its ugly head. What happens if hypertension goes untreated? A blood vessel pops in your brain, causing a massive stroke, or your heart enlarges and thickens—meaning it can't pump blood effectively—or your kidneys don't get enough blood to filter out poisons. You're probably aware of the worldwide epidemic of Type 2 diabetes, which kills by thickening and clogging arteries, leading to heart disease and deadening the nerves to the legs, which can ultimately require amputation. All are great ways to polish off those who take more than their fair share.

Diabetes is just the beginning. How about arthritis? Easy. If it hurts to walk, you won't go get food, or you can't run to avoid being someone else's food. Cancer? In a macabre way, having your own cells eat you alive is a fast way to exit life. Is it any wonder, then, that recent findings show that obesity increases the risk of cancer up to tenfold over that of normal-weight individuals?[20,21] Or that cancer and nerve disease rates are unusually high in extreme athletes?[22-25] Or that cancer is a good predictor of subsequent heart disease or stroke?[26,27] Why? Because once killer genes have been turned on, if they can't get you one way, they'll find another. Have you had friends or family members who survived heart surgery only to later die from, say, colon cancer? Bad luck? No. Killer genes? Yes. Like Clint Eastwood in *Dirty Harry*, killer genes keep firing off bullets until one of them gets you.

The leading cause of death for Alzheimer sufferers is starvation: these unfortunate folks forget to eat. Is this just a coincidence or part of a genetic plan? Exposure to small amounts of sugar actually improves short-term memory, probably so you can remember where you found that new fruit tree.[28] On the other hand, chronic exposure to sugar kills off short-term memory cells. If you're abusing the privilege, what better way to get you to forget where the sugar bowl is? Scary, yes, but a beautiful design for controlling populations.

How about chronic stress? Being at the bottom of the pecking order means an animal gets the worst food, the most unwanted attention from more powerful rivals, the least chance to mate, and generally the highest stress level. These animals usually die early, many from disease, so their mixture of genes is not perpetuated. Low-ranking baboons have been found to have the highest parasite load compared to others in their troop, and are always sick.[29] Runners training for a marathon complain of constant sore throats and colds; their immune system has been impaired, just as a low-ranking baboon's has.

Observe animals under stress and you'll see that they eat to soothe themselves, and I'll bet you do, too. To quickly induce obesity in a rat, repeatedly pinch its tail. The feeding frenzy lasts as long as the pinching continues. Why? Remember the second and third rules: avoid pain and find pleasure. The grain in the rat chow rapidly breaks down into sugar, stimulating the pleasure center; but stop eating, and the sugar load drops and the pain resurfaces. The sugar doesn't eliminate the pain, but it does make it tolerable— although the more sugar the brain is exposed to, the more it takes to control pain. Moreover, stress also activates the killer genes. After all, a stressed-out rat doesn't have a good combination of genes to keep around. The next time you're feeling stressed out, remember those rats: you, too, will be ushered out of the gene pool if you eat the way they do.

AMBER ALERT

Your body has given many of you a great gift in the form of warnings such as heartburn, headaches, joint aches/arthritis, high blood pressure, depression, or elevated blood sugar. Rather than correct the underlying problem, too many of us have resorted to treating them with over-the-counter or prescription drugs. I was trained to believe this was the appropriate course of action as well. But when a warning light on your dashboard alerts you to immediately take your car to your dealership, does the dealer cover up the warning light with duct tape so it doesn't bother you? That sadly is exactly what many of us doctors, myself included, have been doing: covering up the warning signs. If you take medications for high blood pressure, diabetes, arthritis, cholesterol, depression, stomach acid, or pain—or get stents or bypass surgery or have colon polyps removed—you're doing just that: covering up the signals that killer genes have been activated.

Shockingly, the number of people on Medicare being treated simultaneously for four or more medical conditions has doubled from 1987 to 2002.[30] During this same period, the rate of obesity in the United States also doubled—a one to one correlation.[31] Another study of 500,000 "healthy" nonsmoking people aged 50 to 70 over a ten-year period recently found a direct correlation between the degree of overweight and early death, with the more excess pounds, the sooner the grim reaper appears. That's right: That internal calorie counter we all carry in our genetic programming activates the killer genes. In other words, death and disease are your genes' way of saying you're being voted off the island.

I WAS NOT A KEEPER

You may now be beginning to suspect why elite athletes often die young. Why, despite running twenty miles a week and putting in an hour at the gym each day, did I still weigh in at 228 pounds? And why was I still plagued with high cholesterol, high blood pressure, high blood sugar, and frequent colds? Because my activity level, food consumption, and "pain" all conspired to inform my genes that I wasn't a very successful animal, which you'll recall activates killer genes, as does taking more than your fair share of food.

Numerous studies on animals and humans confirm these results.[18,20,25,29-34] As we leave this chapter, it's important to remember two things: First, despite my jokes to the contrary, your genes have no malicious intent. Rather, they're part of a complex communications system, or autopilot, that runs your cellular functioning and body. Second, you're an active participant in the conversation. When you eat and live a certain way, you unwittingly send messages to your genes saying that you're a drain on the system.

Wow! Killer genes, genetic programming, hardwiring to get you to produce copies of your genes? Okay, Dr. G., you've got my attention, but what does this have to do with losing weight? In a word, everything. Turn the page and keep reading.

Chapter 2

WE ARE WHAT WE EAT

YOU now understand why the typical Western diet is the best thing that's ever happened to your genes, even if it eventually damages your own health. Despite the dramatic changes that have occurred in our diet in the last century, the way we eat today is a mere blink of the eye in human evolution. If you've ever seen photographs of your great grandparents, I'm willing to bet that almost all of them were slim. Within two or three generations, Westerners have gone from dealing with the occurrence of starvation for a significant part of the population to experiencing obesity as a serious public health issue.

Over the past thirty years I have seen changes that scare me to death. In fact, I can even look at you fully clothed and pretty much know what your blood vessels, heart, liver, prostate, and brain—yes, your brain—look like. A huge number of you have allowed a condition to occur that guarantees the activation of killer genes.

If you're still in denial, just look at your belly in the mirror. If your gut obscures your privates when you stand up straight and look down, you've already triggered killer genes that have started the process of finishing you off. If your gut overlaps your belt, I can virtually guarantee they've been activated. In a few of you, the signs are too small to see from the outside, but send your genes the wrong messages, and you'll assuredly see them soon.

I was certainly beginning to look like a Pillsbury Doughboy myself, but it wasn't until I turned 50 that I finally came to grips with a startling fact: The people I was operating on for heart disease looked very much like me. We consider any male patient who weighs in at 220 pounds a member of the "Big Boy Club." Every day, I would gaze down at the next person I was about to help, all the while a card-carrying member of the club myself. But no one ever knew. It's amazing how many pounds an Armani suit can hide, even in a size 46.

How could I have let this happen to me? I had studied the influence of nutrition on animals and people, and supposedly I knew how serious the effects of diet and obesity are on the human body. To understand how we as a society have come to this sorry pass, I want to take you briefly through the evolution of the human diet—and also garner some lessons from the carnivores and herbivores of the animal kingdom, as well as from plants.

PLANTS AND PRIMATES

As you probably know, we share 98 to 99 percent of our DNA with chimpanzees and gorillas. So despite our recent dietary habits—like the time you made a monkey of yourself at the all-you-can-eat buffet in Vegas—our genes and our fuel needs are most like those of the great apes. They spend time foraging for and eating leaves and fruit, although perhaps 10 percent of their diet comprises ants, grubs, other insects, and even smaller mammals. The hallmark of an ape's diet, and of any leaf- or fruit-based diet, is consumption of micronutrient-dense but calorie-scant food. And, if given the opportunity, all apes will gorge and fatten up on fruit and ignore other foods during certain seasons.[1,2] Remember this fact.

For years, most of us have associated nutrients with calories. In fact, until the last few years, nutrition training in medical schools preached that a calorie is a calorie is a calorie, no matter where it comes from. End of lecture, class dismissed. Unfortunately, as you'll soon discover, calories merely tell us how much energy is contained in a food, not how much energy your body can derive from it. Nor does calorie count tell us anything about the host of micronutrients a food contains—or their effect on the genetic code in each of our cells, which we'll explore in Chapter 3.

Micronutrients are basically the vitamins, minerals, trace elements, phytochemicals (*phyto* means "plant"), and thousands of other compounds that naturally occur in plants. Micronutrients produce a host of biologic effects in animals that eat them, and actually tell our genes to turn on or off, and to make proteins, fats, and hormones that affect every one of our cells. As you will soon learn, they're key to turning your health around.

But back to apes. The average mountain gorilla has to eat about 16 pounds of leaves a day to get enough calories to maintain its weight.[3] But it's not hard work to get those calories: the gorilla just pulls up a branch and starts munching. In fact, a key distinction between carnivores and herbivores is the pursuit of calories. Unlike herbivores, carnivores generally spend little time eating and move around only to catch food or find new sources of food. Similarly, to avoid wasting their efforts on chasing the fastest, largest prey, lions and other carnivores target the injured, the sick, the old, the young, and the distracted. If a big cat realizes it cannot quickly run down its prey, it breaks off the chase to conserve energy. Despite their very different lifestyles, both carnivores and herbivores expend very little energy overall. Remember, populations flourish when the greatest number of calories can be obtained with the least effort.

What do flesh eaters such as wolves, lions, coyotes, and, of course, dogs and cats do most of the time? That's right, they sleep. They can enjoy their lives of leisure because they're designed to eat calorie-dense foods. When they do eat, carnivores devour their food quickly, rather like some 13-year-old boys I have known. Relative to plant foods of comparable size and weight, animal organs and muscles are packed with calories. Eat one animal and your caloric needs are met for a while.

But carnivores pay a price for the privilege of eating meat. Digesting meat results in a high metabolic rate, so they need to reduce it—or suffer the consequences—which is why they sleep a lot. Animal protein is actually a terrible source of usable calories because the process of breaking it down into molecules your body can use produces heat that wastes about 30 percent of the calories. (In Part Two, you'll learn how higher heat production contributes to rapid and unnecessary aging. We'll come back to how these concepts can be used to help you raise your metabolism in Phase 1, and then we'll shift to methods to lower your metabolism for the long run in Phase 2. The entire premise of Diet Evolution centers on the fact that you will use one technique initially to lose weight and then evolve to another to encourage longevity.)

One more important point about true carnivores: they generally get their micronutrients by eating herbivores, which ate growing plants. Remember this concept. I forgot it; you may never have known it; and giant food companies don't want you to know about it.

EVOLUTION OF THE HUMAN DIET

The fossil record is undeniable: our distant ancestors added meat to their diet of leaves and fruit somewhere in the neighborhood of 2.6 million years ago. Just as we *evolved* as a species, so our diet *evolved* in response to social, climatic, and food-source changes. But let me assure you, this book is not about another "caveman" or Paleolithic diet. Remember, carnivores, because they eat meat, have shorter lives than herbivores. My goal is for all of us to survive—and thrive—for a good long time.

Nevertheless, our species clearly flourished on this new, more carnivorous diet. You don't take over the world without being healthy, strong, and endowed with intelligence. The combination of meat and plants (the so-called hunter-gatherer diet) obviously supplied the proper combination of calories and micronutrients, as the fossil record displays a sturdier bone structure and taller stature in our hunter-gatherer ancestors than in earlier specimens. Adding micronutrient- and calorie-dense animal food also gave early humans more free time than other apes.

One more thing: wild carnivores obtain their plant micronutrients by eating the herbivores that ate the plants; and until about fifty years ago, humans also got many of their micronutrients from the meat of animals that grazed on grass and other plants. Humans even lost the ability to manufacture vitamin B_{12}. Huh? What does that have to do with eating a rare hamburger? Well, if a vitamin necessary for life is not found in abundance in the diet, an organism can usually manufacture it out of molecular building blocks. Because cows and other animals, but not plants, make B_{12} in large quantities, we get it in our diet and no longer need to make our own. Likewise, most animals manufacture their own vitamin C, but humans don't. We can surmise that plants high in vitamin C were so abundant in our early diet that we simply lost the need for that genetic code—or, more accurately, it is just not "turned on." Why give up a potentially useful gene sequence? Because, from a biological point of view, it's excess baggage. It also takes energy to manufacture anything. Interaction with the new animal foods altered our forbears', and consequently our own, genetic ability to manufacture a vitamin necessary for life. Pretty powerful stuff: something you eat a lot of can control or change your genes!

As the chart "Evolution of the Human Diet" on the opposite page demonstrates, our diet has evolved from that of early humans, to those of hunter-gatherers, to the diet resulting from the Agricultural Revolution, and finally to our modern diet. In that time, calorie-dense and micronutrient-sparse food has supplanted micronutrient-dense and calorie-sparse food. Study it well because, in Diet Evolution, you'll trace this evolution gradually backward, one step at a time. You'll see some interesting patterns emerge. Early human's diet was densest in micronutrients but sparsest in calories. Remember, leaves

EVOLUTION OF THE HUMAN DIET

How our eating habits have changed over the last million years

EARLY MAN [*Pleistocene Era*]

40%	10%	50%
Protein from herbivores & vegetables	Plant & animal fats	Leaf- & fruit-based carbohydrates [no grains]

HUNTER-GATHERER [*150,000 Years Ago*]

40%	10%	50%
Protein from herbivores & vegetables [but no legumes]	Plant & animal fats	Leaf- & fruit-based carbohydrates [no grains]

AGRICULTURAL REVOLUTION [*10,000 Years Ago*]

30%	25%	25%	20%
Protein from legumes & grazing animals	Animal fats	Whole & stone-gound grains	Vegetables, tubers & roots, dried & occasional fresh fruit

MODERN ERA [*From 1900 to Today*]

15%	25%	30%	25%	5%
Processed foods*	Protein from animals fed grain & legumes	Animal fats & grain-based oils	Refined carbohydrates & sugars, fruit & fruit juice	Vegetables

* What I call "white," "beige," and "brown" foods. See pages 66–69.

have very few calories for their volume. Our modern diet is the exact opposite. In the last half-century, not only have we dramatically changed our diet to include lots of grain (most of it refined), ever more meat from animals raised on yet more grain, and endless amounts of sugar, we have also reduced our consumption of plants full of micronutrients. Contrast that with the Agricultural Revolution diet, which persisted until about a century ago, when the animals we ate were rich in micronutrients because they grazed on grasslands and prairie.

CALORIES VS. VOLUME

The *volume* of food that humans eat, on the other hand, has held constant throughout time.[4] So it stands to reason, as the chart on page 27 illustrates, that we have consumed more calories at each point of our dietary evolution. Make sure you understand this point: you're not eating more food; you're eating food that has more calories per cubic inch. In fact, most of the food you consume today didn't even exist a hundred years ago. Is it any wonder that even as our genes are as happy as clams, our bodies are crying out by manifesting their distress in the form of obesity and a long laundry list of chronic conditions?

Until very recently, our intestines—and our genes—interacted primarily with green plants and the compounds they contain. Don't be confused by this. While it is true that early humans supplemented their diets with animal meat and fat,[4,5] there is no evidence that animals ever became the *dominant* food source. Even if early humans got most of their calories from animal products, day to day, they relied primarily on plant material to fill their stomachs.

ENTER AGRICULTURE

The Agricultural Revolution occurred roughly 10,000 years ago, probably as a result of overhunting, which made domestication of large animals the only way to ensure enough calorie-dense food. Agriculture also has time- and energy-saving advantages. It's a lot faster to walk out your back door and pick up some eggs, milk a goat, or kill a sheep than to have to hunt for your food. Raising animals was a biggie, but the domestication of grains was the real turning point in population growth and social history. Now there was micronutrient-dense, calorie-dense plant food that humans could cultivate when and where they wished, as well as store and transport. Grains allowed humans to survive and thrive. More humans, more genetic copies; your genes were and continue to be delighted with this turn of events.

But not so fast. Planting, cultivating, harvesting, and preparing grains without machinery require a massive expenditure of energy. Unlike with primate or carnivore models, producing this new high-calorie diet burned off a lot of calories. Rule number 1 of survival is to find the most calories for the effort; rule number 2, its corollary, is that the less work you have to expend to get calories, the better. Human beings' early agricultural lifestyle was a remarkably good match for rule number 1, but a lousy match for rule number 2.

It was this mismatch of agriculture and energy use that sets us up for disaster in the twentieth and twenty-first centuries. Until very recently, your forebears expended huge amounts of energy to obtain high-calorie grain products; now you can get them with little or no effort. Your genes couldn't be happier, but you are paying the price. Even the Egyptian pharaohs weren't immune to this rule: the recent discovery of the only female pharaoh revealed that she died from diabetes. She was a grain eater who didn't labor for her dinner. Sound familiar? Three thousand years later, the same overuse of the wrong types of grain products, even of whole grains, is largely to blame for much of our society's entwined weight and health problems.

What do I mean by that? As you'll soon see, if calories consumed are higher than energy expended, ill effects occur, because you are perceived by your computer program as taking more of your share, thereby activating killer genes. We are hardwired to find the most calories for the least effort, but in any ecosystem there are only so many calories to go around. But why then do we seem to just keep eating? I think our genes are looking for something: the building blocks of cells, the micronutrients. So important are micronutrients to proper cellular functioning that a few of us in the research community believe that our bodies are genetically programmed to continue eating until we consume a bare minimum of micronutrients. If these are not present, we keep eating, assuming the next bit of food will contain the micronutrients we need not just to survive but also to flourish.

LESSONS IN THE WILD

With the exception of animals that hibernate or anticipate periods of starvation during the winter, there simply are no overweight wild animals. An individual animal's genetic code won't allow obesity to occur because, when adequate stores of fat are present, a sophisticated system of monitoring cells produce hormones that shut off hunger signals. Also, as just discussed, quite possibly the micronutrients in the foods in an animal's habitat act as a satiety switch. Change in the amount of daylight is another likely satiety switch. Later you will learn how to use both of these satiety switches to help moderate your own appetite. We also now know that fat cells, and the other cells that surround them, produce hormones and other potent substances that convince animals to stop eating. Little did you know it, but you are awash in these substances. But food and pharmaceutical manufacturers provide you with plenty of ways to ignore these warnings.

CALORIC TRADEOFFS

When primates share an environment with predators, as on the African savanna, they tend to form groups composed of several large males for defense and a lot of smaller females to bear offspring. Where these circumstances don't exist, male and female monkeys are the same size.[6,7] Supporting muscles requires lots more calories, which are hard to come by on the plains, so it doesn't make evolutionary sense to make everyone big and muscular to defend the group against predators. A few big linebackers are all you can afford. Producing offspring also requires additional calories, so small females with lower calorie needs result in more babies. Whenever predator pressure or struggles to survive occur, the genetic sequences that promote early puberty and frequent pregnancies are activated.

Our ancestors were also programmed to deal with food scarcity during droughts, rainy seasons, and winter. These cycles were so frequent and consistent that humans—and all other animals—developed genetic sequences that activate only during periods of starvation, protecting cells from damage. Obviously, if a creature makes it through a period of starvation, it can pass along its genes. In Phase 3 of Diet Evolution, you'll learn how to activate these genes—without the starvation!

PLANTS AS ALCHEMISTS

Now let's learn some lessons from plants. Just because a plant produces juicy fruit that we enjoy eating, that doesn't mean that the plant "thought" up the idea as a way of getting its seeds distributed. Or does it? And how about using substances that affect our brains to get us to do their bidding? Please, that's just ridiculous! Or is it?

Plants have been around a lot longer than animals, but when animals, particularly insects that feed on plants—let's call them "plant predators"—arrived, it was a whole new scene. Since a plant can't run, hide, or fight back, what defensive strategies can it use to survive? As alchemists of the highest order—able to convert sunlight into matter, no less—plants have developed sophisticated systems to produce biochemicals out of light and basic nutrients such as carbon dioxide, water, and minerals. In their efforts to pass their genes on to the next generation, plants have developed thousands of compounds to convince animals, particularly insects, not to eat them.

Seeds carry the next generation of a plant's genetic code, and the mature leaves produce the energy necessary to generate offspring, so plants put most of their defenses into seeds and mature leaves as the storehouses of animal toxins. Herbivores learn this, preferring to nibble on young leaves. That's why caterpillars hatch just as trees and plants leaf out. Ever notice that your dog loves to nibble the fine shoots of early grass, but eats mature grass only if it wants to vomit? Now you know why.

The knowledge that plants concentrate most of their anti-animal defensive systems in their leaves and seeds is going to have profound implications for us as you move through Diet Evolution; indeed, it is the key to Phase 3. At this point, just remember that the longer humans have been exposed to a plant-based food, the longer our genetic code has had to come up with a defense strategy. Detoxification enzymes, which neutralize or eliminate toxins, are located mainly in the liver and are usually on standby in the event of an emergency. For grazing animals, the odds are that they are not going to eat enough of any single semi-toxic food to chronically overwhelm their enzyme systems. If one plant doesn't agree with them, most animals naturally just move on to another kind.

Raw vegetables are loaded with these "toxins." Am I saying you shouldn't eat plenty of raw vegetables? No way. But you do need to *evolve* your eating pattern to allow your liver enzymes to handle the additional load of plant toxins. Do this gradually, as regular exposure to these toxins will activate the liver's detoxification systems, so you can get the benefits of eating lots of vegetables without the negative effects. You'll learn how to do just this in Phase 2.

SWEET MANIPULATION

Most plants with indigestible seeds use an entirely different strategy for survival. The plant *wants* you to eat its seeds so they will emerge whole at the other end of your intestinal tract, in a new location, surrounded by a generous dollop of fertilizer. The plant has a vested interest in first attracting its predator's attention and then convincing it to eat as much as possible, which it does by manipulating your genes and, in turn, your hormones. This strategy is diabolically effective, with stunning implications for your health.

Fructose, or fruit sugar, is unique in many ways. Ordinarily, sugar (glucose) is absorbed directly from your intestines and enters the bloodstream. Eat a significant amount of sugar, and your pancreas secretes insulin—a growth hormone that tells your body to store sugar as fat—into your bloodstream. As insulin levels rise, a feedback system in the satiety center in your brain gets activated. Additionally, the presence of glucose in your bloodstream releases a fat-regulating hormone called leptin, which signals your brain that you've had enough to eat. So glucose transported in your bloodstream gives your brain a double dose of hormones, telling you to stop eating.

How does a plant override an animal's "I'm full" switch? It produces fructose, another form of sugar, which circumvents the activation of both these warning signals. Rather than enter the bloodstream, fructose goes directly to the liver, where among other things it stimulates the formation of triglycerides, the precursor of what we commonly refer to as cholesterol. (Remember this fact about triglycerides; it's a key to success later.) Because fructose doesn't enter the bloodstream, insulin rises much more slowly and leptin isn't activated. Sneaky fructose bypasses both feedback defense mechanisms your body has developed to stop overeating.

TOO MUCH OF A GOOD THING

"Tricked" or not, fruit-eating animals clearly benefited from this relationship. Because fructose doesn't trigger satiety signals, the animals could consume far more calorie-laden fruit than they would if its sugars took the normal pathways in the body, so more calories were stored as fat. Conveniently, most fruit ripens in late summer and early fall, shortly before food stores traditionally become scarce. Even in the tropics, fruit ripens seasonally. That's why, for example, orangutans in Borneo only store fat during fruiting season, prior to the dry season when food is less available.[1] Historically, animals that stored sugar as fat made it through the winter and were more likely to reproduce come spring; those that didn't died off. So the fundamental program that runs your genetic computer is simple: If a food tastes sweet, eat lots of it because winter is coming and you need all the body fat you can get to survive until

Insulin Works Overtime

. . .

Insulin has three jobs:

- It delivers sugar and other calories to cells for immediate energy production.
- It tells the liver to convert excess sugar into fat for long-term energy storage.
- It stimulates cells to grow.

Insulin gets us in trouble with its second function. The only thing that would make our early ancestors' insulin levels rise rapidly was a ton of fruit sugar, unlike leaves, tubers, nuts, or meat. Their genetic autopilots would correctly assume that eating lots of fruit meant it was late summer or early fall, and release of insulin would signal their liver to turn sugar into fat so they could make it through the winter. The higher your resting insulin level, the more your liver "thinks" that winter is coming and therefore it better keep churning out more fat. Our Western diet is so full of sugars and refined grains, which behave exactly like sugars, that most of us are sending out this message 24/7. In fact, any sweet substance, *even an artificial sweetener,* tells your computer program that sugar is on the way, so please produce insulin. No wonder no study has ever shown that sugar substitutes have any benefit for weight loss, and that even just the *taste* of sweetness raises insulin levels.[8–11]

spring. I call this system "Store Fat for Winter." This system works great when winter food supplies are scant, but it's a recipe for disaster when food is constantly available, as in our society.

Why doesn't our autopilot stop this obviously life-threatening process? By now, you should know the answer: sweet foods and syrupy sodas provide mucho calories for minimal effort. They'll help a girl mature faster so she can reproduce sooner. Her autopilot doesn't care what it might be doing to her arteries. Studies of infants who were nursed long term (four years) show fatty cholesterol deposits in their arteries at an early age—breast milk is loaded with the milk sugar lactose.[12] I've seen these plaques in many youngsters on whom I've operated.

. . .

Also known as the metabolic syndrome, or syndrome X, insulin resistance afflicts more than 60 percent of Americans, most of them unwittingly, and is a manifestation of killer-gene activity. When you have a lot of muscle cells, as you do in your youth, the insulin response usually functions efficiently, but when you lose muscle mass and pack lots of fat cells in between blood vessels and muscle cells, insulin has to push sugar through mounds of fat to reach your muscles. Then when it finally gets there, the cells of the muscles are already full, so your pancreas does the only thing it knows to do: produce more insulin. As your insulin levels and blood sugar continue to rise, your liver gets the message that you are happily munching on fruit, so winter must be fast approaching, and it obligingly stores fat for the "winter" that never comes. Now you understand why insulin resistance is rampant in our society.

Do you have insulin resistance? In addition to a gut that hangs over your belt or a thickened waist, other sure signs of insulin resistance are rosacea and skin tags. To assess your risk, just remember one of the sayings that my patients call "Gundryisms": *Fat on your ass, you're built to last! Fat in your gut, you're out of luck!* Your fat also manufactures testosterone, so if you are a woman and your hair is thinning, you also probably have insulin resistance. Want some motivation to get your hair back? Dr. G. says: *If you lose your gut, you'll find your hair!*

THE ANTI-NUTRIENTS

What defensive mechanisms do plants without protective casings, toxic substances, or sweet fruit have to protect their seeds from plant predators and ensure their survival? Enter the anti-nutrients. Most legume (bean) seeds, for example, contain compounds called phytates, which effectively slow or prevent the absorption of vitamins, minerals, and other nutrients from the intestinal tract of animals. So the more of certain seeds an animal eats, the less nourishment it receives. You can bet that an animal losing weight rapidly learns which plant parts to avoid. (Cooking beans diminishes but does not eliminate the effect of phytates.)

Another type of anti-nutrient acts in a way similar to that used by cancer specialists to destroy or slow the development of rapidly growing cells. Some plants produce "fake" compounds that molecularly mimic important proteins or enzymes essential for cells to grow and divide. If an insect eats these "fake" proteins, they get incorporated into its cells and prevent growth and division. Bye-bye, insect. But before you shrink from ever eating another leaf of lettuce, remember that humankind has evolved alongside plants over millions of years, during which we have relied on them for sustenance. We and other animals have learned to consume these compounds, which actually act as mini-doses of chemotherapy, preventing ever-arising and ultra-fast dividing cells from growing.[13,14] Chronic exposure to lots of plant compounds also appears to produce a low level of toxemia that actually activates an individual's longevity genes. What I'm suggesting is that many vegetables are "good" for you because they're "bad" for you! You will learn how to harness the power of this exciting discovery, called hormesis, in Phase 3 of Diet Evolution.

THE VEGETABLES VS. MEAT CONTROVERSY

In societies where children consume large amounts of vegetables, they tend to be shorter in stature as adults and begin reproducing later in life than those who don't rely as much on vegetables. But when these populations switch to more animal and refined-grain sources of food, growth rates and stature increase. Witness how in just one generation after the introduction of the Western diet, the Japanese have become significantly taller.[15] Faster growth, greater height: must be better nutrition, right? Not so fast. Ask the petit Okinowans or the short Sardinians, both among the longest living people on earth, if short stature signifies poor nutrition and a shortened life span. On the contrary, taller individuals tend to die earlier than their shorter peers.[16]

Now let's sort out the decades-long debate between the animal protein diet and vegetarian diet gurus, with their conflicting "good foods/bad foods" lists. Both camps see the tree but miss the forest. What do I mean by that? The high-protein group points to all the anti-nutrients in grains, seeds, and beans, and proclaims how all this stuff will kill you. But as you're starting to discover, when taken in the correct dose, the anti-nutrients in plants activate the hormesis response, which prolongs your life with low doses of poisons that effectively tell your genes to protect you from this threat. So vegetables are good for you because they're "bad" for you.

So, the vegetarians are right? It's not that simple. Almost all so-called herbivores, including our nearest relative the gorilla, consume some animals or animal products in the form of insects, grubs, worms, and such that happen to be on the leaves the herbivores are munching.[3] Marmoset monkeys, which eat nothing but fruit, are unable to reproduce in zoos unless their diet is fortified with 6 percent animal protein. It turned out that the "perfect" zoo fruit was missing the worms and bugs naturally present in fruit in the wild![17] So it would appear that some animal protein is a good thing.

However, eating animal protein, especially muscle, which is high in iron, is a double-edged sword. Breaking protein down into usable fuel in the digestive tract liberates an excessive amount of heat, which induces rapid aging. (Remember those dozing carnivores?) This excess heat production can be used to your advantage during initial weight-loss efforts; in fact, it is the secret of success in high-protein/low-carb diets. But in terms of longevity, the iron overload in animal protein induces oxidative damage to your tissues. Moreover, in terms of weight loss, animal tissue or products—think cheese—have dense caloric loads that thwart long-term weight-loss efforts. Most of the low-carb dieters I have treated reached a plateau fairly quickly, from which they could not advance without reducing their intake of animal foods. Diet Evolution finesses this problem, as you'll soon learn, and so resolves the conflict between doctors Atkins and Ornish.

ACCOUNTING FOR TASTES

If you find yourself irresistibly attracted to foods that are either sweet or salty, you're not alone. You're actually hardwired to seek out these tastes and fat because, in times of scarcity, they would have enhanced the likelihood of passing on your genes to your children. Sweetness, salt, and fat are each essential for survival, but in excess each has become a hindrance to our health.

SWEET DECEPTION

Sugar is a generic term for simple carbohydrate molecules that are the main energy-storage molecules for plants. Link one or more sugar molecules and you get a form of sugar once referred to as starch, but usually now called a *complex carbohydrate.* Rice, wheat, potatoes, and beets are just some complex carbs that break down into sugar when they're digested. In plants, sugars usually appear in concentrated amounts in either the seeds or fruits that

A Bitter Warning

. . .

The taste buds that identify bitter tastes occupy much less territory than those for sweet, but are strategically located near the tip of your tongue to act as an early warning system. Plants that make particularly lethal compounds usually formulate them in bitter-tasting vehicles to warn animals of the consequences of consuming them. The plant would rather you take a nibble and stop eating than have you eat it and kill you both. Studies have shown that your autopilot is activated the minute a bitter taste hits your tongue, causing an immediate rejection much like that when you touch a hot stove. But as Nietzsche famously said, "that which does not kill us makes us stronger." Numerous societies notable for their longevity have a fondness of bitter greens and other foods.[18,19] That's why I say: *More bitter, more better!*

surround them, or in the roots. Sugars in the seeds mainly feed the growing plant embryo until it can start manufacturing its own food from sunshine; those in the fruit attract animals, which carry the seeds off to greener pastures. As you've learned, fruit sugar, or fructose, is specially designed to make unwitting animals eat copious amounts.

But what's in it for the animal? As it turns out, a lot! It's in your genes' interest for you to eat the sweetest food with the most calories. Just in case you have a different opinion, your genes have a powerful ally in the tip of your tongue, which houses your taste buds. Sugar is so powerful that your whole tongue is oriented to finding it, and your brain is programmed to seek it endlessly, because if it gets in your system fast enough, it goes right to your brain's pleasure center. In the natural world, the only source of concentrated sugar with enough power to elevate your blood sugar level is a lot of fruit. Unfortunately for you and me, that wasn't good enough for our food scientists, who have concocted all sorts of products masquerading as foods that leave natural sugars in the dust. And your genes love them. In Phase 1 of Diet Evolution, you'll learn how to lessen the power sugar holds over your genetic program and how to convince your autopilot that winter is not just around the corner. To help you with this concept, it's worth memorizing: *If it tastes sweet, retreat!*

CHEWING THE FAT

We are also programmed to find fat, which enhances the flavor of foods (making it so irresistible), and carries more than twice the number of calories by weight than sugar or protein. At what time of year would animals have the highest fat content in their bodies? In the late summer and early fall, after a season of grazing, getting ready for winter, of course. The combination of large amounts of sugar and fat in any food triggers the same "Store Fat for Winter" program. No wonder you can't resist this combo. Your genes are looking for sugar and fat to signal an eating binge before winter sets in.

Fat has another bonus going for it. With the exception of vitamin D, which your body can manufacture directly from sunlight, you must get the A, E, K, and other fat-soluble vitamins in your diet, and they must be consumed as or with fat. The fats that we used to consume as recently as fifty years ago came from animals that fed primarily on grass and leaves. Remember that carnivores get most of their micronutrients by consuming the fatty organs of their grass-eating prey, a fact critical to understanding this second innate drive to find fat. It would have only been in this fat that our ancestors obtained the fat-soluble vitamins necessary for their health. (This is why later you will see that I recommend eating beef that has grazed on grass, not been fed corn in a feedlot.) Moreover, this is where they also obtained the essential omega-3 and omega-6 fats.

ESSENTIAL FATTY ACIDS

You have probably heard of omega-3 fats as the essential fatty acids present in such coldwater fish as mackerel, sardines, bluefish, and wild salmon. These fish obtain their omega-3s from eating algae, so they act as a middleman between you and the algae, much as cows do between the micronutrients in grass and you. Grains and some seeds contain large amounts of omega-6 fats: feed a cow—or you—grain, and the result is more omega-6s in its—or your—fat. The term *essential* means that your body cannot manufacture these fatty acids and must get them from food, but the term also is apt because about 70 percent of the insulation system for the nerves in your brain is composed of the two most common omega-3 fats, docosahexanoic acid (DHA) and eicosapentaendic acid (EPA). The more depressed a person is, the lower his or her level of omega-3s.[20] In fact, institutionalized patients given heavy doses of omega-3s resolve their depression better than those given antidepressants.[21]

Omega-3 and omega-6 fatty acids are also the main building blocks of our hormone systems: omega-3s are used to construct anti-inflammatory hormones like prostacyclins and prostaglandins, while omega-6s are used to manufacture inflammatory hormones, such as arachadonic acid and thromboxanes. You need both types to have a proper communication system within your body. In a traditional ape's or foraging human's diet, a ratio of omega-3 to omega-6 fats of between 1:1 and 1:2 appears to be critical in balancing inflammation and anti-inflammation hormones.

With the Agricultural Revolution, the grain-based diet emerged—although fat and protein still came from animals grazed on grass—and the ratio shifted to 1:4, which accounts for the presence of arthritis found in the joints of Egyptian mummies. Today, the ratio is between 1:20 and 1:40. This staggering increase results from the declining intake of vegetables and grass-fed meat and the concurrent explosion in the use of corn oil and other grain-based oils and products whose fats are primarily omega-6 based. Even the fats in beef and chicken, which used to be heavily omega-3 based, are now in the omega-6 camp, thanks to feeding our livestock grain and soybeans. Today's cow is not the same cow of fifty years ago. Ditto for chickens.

Does your ratio of omega-3s to omega-6s really matter? Take it from ground squirrels. They won't go into hibernation if the ratio is disrupted in the laboratory.[22] Their genes "know" something is amiss and won't risk it!

INFLAMMATORY DISEASES ON THE RISE

With our production of inflammatory hormones now forty times greater than the anti-inflammatory ones, what diseases and conditions would you expect to appear? How about arthritis, asthma, skin lesions like psoriasis, eczema, and autoimmune diseases like lupus, multiple sclerosis, and Crohn's? Ironically, we're looking for the sinister cause of the staggering increase in childhood asthma, all the while stuffing our youngsters with refined grain products. Is it any wonder that we're depressed, achy, and having our hips and knees replaced right and left?

But, evidence is mounting that high-dose omega-3 therapy is effective in reversing many of these inflammatory conditions. What's surprising is that we are only now rediscovering these facts. More than a hundred years ago, cod liver oil was introduced into Europe and textbooks of the time are full of cases of patients with crippling rheumatism (arthritis) who recovered once they partook of fish oil. Why the rheumatism? The Europeans loved their bread, with its omega-6 profile.

SALT OF THE EARTH

Let me introduce you to a new way of looking at the body you call home. From now on, I'd like you to think of yourself as a portable saltwater aquarium tank. I'm sure you're ready to commit me by now, or at least get professional help for yourself for having bought this book, but please bear with me a little longer. As you probably learned in biology class, you and all other terrestrial animals by weight are 70 percent salt water. Even many of the trace elements in our cells and blood are in exactly the same concentrations as found in our oceans. If the concentration of salt gets too low in our blood or cells, the delicate cellular machinery sputters to a stop, as surely as if you ran out of air underwater. (Most deaths suffered during marathons occur not from heart attacks but from drinking too much water and losing too much salt.)

Unfortunately, your body has a built-in salt leak. Every time you get hot and need to cool down, you toss off salt water as sweat. As a result of this poor design, you're programmed to constantly be on the lookout for salt and fill up when you find it. After all, salt is a rare commodity away from the oceans. Whenever I return from a run with my three dogs, the first thing they do is lick my legs. No, not because they like me or I smell particularly "ripe," but because they're after the salt in my sweat. They have to replenish their "tank" every chance they get as well. Deer and cattle stampede for salt licks out on the range. From your genes' standpoint, salt is good, and the odds are that you will eat more of everything to which salt has been added. But, of course, too much salt is not good for *you.* Our genes could never have foreseen the amount of salt in the modern diet. So: *Halt if you taste salt.*

HARDWIRED TO EAT JUNK FOOD

In the course of my research, I discovered why so many of us—myself included—have gotten ourselves into this unhealthy mess. Very simply, our genes, our instincts, our programming direct us, just as they did our ancestors, toward sugar, fat, and salt. But before you jump up and down about how you just knew that pasta, potato chips, Big Macs, ice cream, and cookies were good for you because a deep-seated urge draws you to them, I've got some bad news for you. Your hardwiring is there to make sure you perpetuate the life of your genes and others like them via offspring; it has nothing to do with sustaining or improving your life. Next time you watch the space shuttle launched into orbit, watch what happens to the two booster rockets when their fuel is used up: they're jettisoned! They did their job, now they're excess baggage, dragging down the shuttle with their weight. Off they go! Think of your teenager as the space shuttle and you and your mate as the two booster rockets.

Speaking of teenagers, why are they growing up so fast? Just ask your genes. Early human's diet prompted slow growth and late menarche. Even in 1900, the average age at which a girl began to menstruate was 18.[23] Such a diet benefited an individual's health and longevity, but the overall imperative to reproduce genes rapidly took a back seat. In the great contest between individuals and their genes, individuals were the winners and species the losers. Now consider our high-calorie modern diet, which promotes fast growth and early menarche: girls now mature sexually as early as 8. Pregnancy rates in every society, including the United States, show a direct correlation to the rate of weight gain.[24,25] The modern way of eating is the perfect diet to produce more genetic copies sooner. Now, genes win and individuals lose.

GENES IN THE DRIVER'S SEAT

Until about 100 years ago, the information we'd been feeding our genes in the form of food we consumed had been fairly consistent, with the exception of the relatively "last-minute" (okay, it's been about 10,000 years) addition of grains. In fact, grains represent such a recent change that they likely have introduced nutrients with which our genes had never previously interacted. In 1890, the industrious Swiss invented a steel-roller mill that allowed commercial production of flour minus its oil-rich germ and fiber-rich bran, rendering white flour almost completely devoid of micronutrients. Its introduction fired the first warning shot that manipulating our food supply could have disastrous consequences on our genes and our destiny. Unfortunately, the shot wasn't heard until the 1920s, when diseases caused by vitamin deficiencies, notably of B vitamins, began killing *millions* of Americans whose diet contained a lot of white flour. I repeat, millions! Remember the old rule of twenty years? Its ugly head rears again. Within one generation during which white flour became the standard, disease and death were rampant.

Responding to this newfound threat to public health, the federal government mandated that white flour be fortified with a bare minimum of eight essential vitamins and minerals—the very ones that steel-roller milling removed. To avoid a public relations nightmare, the new flour was called "enriched white flour," marketing-speak for "stuff so lethal that the federal government made us put some other stuff in it so it won't kill you as quickly." That 1920s mandate made white flour "safe" again. We loved our white flour so much that we wanted it at any cost.

Grain products have been a major food source, seemingly without considerable problems for the last 10,000 years, so why did white flour and other highly processed grains initiate the rule of twenty years? Unfortunately, once finely ground and processed, grains can be rapidly digested and assimilated into the bloodstream as sugar—and do so at least as quickly if not more so than table sugar. This phenomenon has prompted a host of glycemic index and glycemic load tables, which quantify how fast a certain food enters the bloodstream relative to a sip of sugar water or a piece of white bread. Just as eating lots of ripe fruit floods your body with sugar, so do refined grains, meaning that they also activate what I have termed the "Store Fat for Winter" genetic program. Dr. Atkins and other low-carbers regard refined carbohydrates as the root of all evil, but they see only the trees, not the forest. Avoiding refined carbohydrates was a good start, but it is just one piece of the puzzle.

Let me end this chapter on a cautionary note. Baboons, which subsist in the wild on leaves, nuts, fruit, insects, and small animals, normally have low cholesterol and no coronary artery disease—which would suggest that they have no genetic propensity to high cholesterol. But when they take up residence on hotel grounds and feed off food tourists have discarded, the baboons' lives change rapidly. Yes, they grow larger and faster, and females become sexually mature at an average age of 2 ½ instead of the usual 5. But the bad news is that males develop high cholesterol and often die young of—you guessed it—coronary artery disease, not to mention other human diseases such as tuberculosis.[26] In one generation, baboons with a lifestyle similar to ours—minimal exercise and a Western diet—mirror the contemporary human condition. Their "good" genes have turned "bad," just as predicted.

In Chapter 3, I'll begin to show you how you can reverse the damage already done to your system and find a new way to communicate with your genes so they work with you instead of against you.

Chapter 3

CHANGING THE MESSAGE

By now it should be crystal clear that we humans are genetically programmed to eat the very foods—those high in sugar, fat, and salt—that make you pork up, threaten your health, and reduce the likelihood that you'll be a robust nonagenarian. The very fact that you're here today means that you come from a long line of ancestors who were very good at converting sugar into fat. As you've learned, for most of us, eating ground-up grain products, drinking beverages or eating foods full of sugar, and even experiencing the taste of sugar in artificial, noncaloric sweeteners convinces our computer program 24 hours a day, 365 days a year, that winter is coming and our body must make fat in order to survive.

You've also learned that you may be saddled with hunger genes, sometimes labeled "thrifty genes" by the popular press. You may even feel powerless to oppose their instructions and blame them for your weight problems, abandoning hope of ever gaining the upper hand. The good news is that you're wrong. Your genes are just following your *current* instructions and will follow new instructions that you'll learn shortly. In Part Two, you'll learn how to stop communicating these incorrect messages to your genes, as the first step to taking control of your health, your weight, and your life.

SAME GENES, NEW CELLS

As I tell my patients, I am the same obese person with the same genetic program who now occupies a thin, fit body—and by the way, I eat more food than I've ever eaten before. But really, I'm not the same person because almost every cell in my body has been replaced with revised, reinvigorated cells, using top-grade building materials. But even though my cells have changed, my genes have not. Rather, the information I feed them has changed. By this I mean that the food I ate and my exercise program when I was obese fed my genes information that kept me at that weight and well on my way to diabetes, heart disease, hypertension, and arthritis. The food I ate to lose weight and continue to eat gives these same genes different information that allowed me to almost effortlessly trim down. My autopilot is still at the controls, but the difference is that now the flight plan I've provided no longer flies directly into a mountain!

I'm sure you have heard that feeding your genes "right" will make everything fine. I used to believe that. But I now realize, and hope you have become convinced, that we have been feeding our genes right for some time (from their point of view), and that's why everything is going wrong.

NOW FOR SOMETHING REALLY DIFFERENT

Unlike most diet programs you've probably followed, on Diet Evolution there are no complex formulas, no counting calories, no need to have 30 percent of this or 40 percent of that, no need to count grams of fat or carbohydrates, no need to distinguish between simple and complex carbs or weigh food, no glycemic index charts or complex formulas to memorize. Rarely will you need to measure something. I'm convinced that's why Diet Evolution works in the long term. Can you imagine the look a 110-year-old Okinawan woman weeding her garden would give you if you asked her about the ratio of simple to complex carbohydrates in her afternoon snack? Or how an elderly Sardinian herding his flock up a rocky hillside would react when told his lunch should contain no more than 4 grams of monounsaturated or polyunsaturated fat? Yet, this is what most diet gurus insist you do. Successful centenarians have proved such complicated guidelines unnecessary.

And come on. How long are you really going to do these things? Our evolutionary diet was never about such controlled perfection. Our genetic program evolved from foraging animals and nomadic ancestors. Our ancestors ate whatever they could find or catch. Nobody looked at a sundial and said, "Oops, excuse me, if I don't eat 7 grams of good carbs and 2 ounces of protein from a lean animal right now, my diet will fail." So, as you look over the list of foods to eat, remember that in Diet Evolution, although you can eat plenty of food, you're going to lose weight without worrying about fat grams, trans fats, satu-

The Rabbi Who Looked Like Santa Claus

. . .

The rabbi was so impressed with how several members of his congregation were doing on Diet Evolution that he came to see me himself. At 63, 5' 8" tall, and weighing 240 pounds with 40 percent body fat, he could have been my twin brother six years earlier. He was on a half-dozen medications to reduce cholesterol, blood pressure, and stomach acid. He was diabetic and already had a stent. Were it not for his religious orientation, he could have passed for a jelly-belly Santa Claus.

The rabbi was to the point: "I want to look like you," he said. His blood test results told the whole story: high blood sugar; very high insulin levels; a great total cholesterol of 133, but his HDL was very low and the dangerous type of LDL particles made up nearly half of all his cholesterol. He was shocked. His regular doctor had assured him that his cholesterol was perfect. I assured him that this was typical of my business, just as it is in his: "Rabbi," I said, "People look, but many do not see what is before them." He started Diet Evolution that day.

That was a year and a half ago. The good rabbi is currently down 72 pounds to 168 with 26 percent body fat. Along the way, we have ceremoniously cut new holes in his belt as his waist shrank from 46 to 32 inches! His blood sugar, inflammation, cholesterol, and blood pressure measurements are now normal and he is off all medications. Now that he has trimmed his beard, he does look like me! His faith in a higher power is stronger than ever: he's convinced that's who created the genetic computer program I taught him about. In any event, we both agree that it's a beautiful design.

rated fats, good carbs, bad carbs, or sneaking books into restaurants to check the calorie count in rutabaga mousse (it's pretty high, actually).

Instead, my program has just a few simple rules. Follow them and I can virtually guarantee you will restore your health and your figure. You'll also be guided by simple rhymes to help you remember what certain foods do to your genetic program and your health. I've already introduced you to a few of what my patients affectionately dubbed "Gundryisms." For example: *If it tastes sweet, retreat!* The only other things you'll need are some simple food lists, a few measuring tips, and—oh, yes—a bathroom scale, preferably one with a

Hundreds of books extol the benefits of fruits and vegetables, exercise, and myriad diets. I'll wager a guess that you've regularly tried new programs, whether low-fat, high-fat, low-calorie, high-carb—you name it, it's out there. Ditto with fitness programs: aerobics, strength training, yada, yada. You were probably able to stick with your new program for a while, but before you know it, you were right back where you started, often with an unwanted gift of upsized love handles or a bigger beer belly. Plus, you beat yourself up for being unable to stick to your guns.

Enough guilt already! I'm here to tell you that the reason we struggle with our weight is not our moral failings, nor because we are carbohydrate addicts, but because of the way our genes are programmed. The myriad diets and lifestyles you have probably tried without lasting success are merely variations on modern food and modern behavior, and are therefore doomed to failure. As either Einstein or Ben Franklin said—the dispute rages on—"The definition of insanity is doing the same thing over and over again, and expecting different results." Instead, what I have to offer you is a new theory based on ancient principles.

body-fat percentage feature. (You'll want to weigh yourself each morning immediately after you wake up and take your bathroom break.)

It's not that all other diet books are wrong. Several describe healthful dietary approaches, but they represent a radical departure from the typical way of eating—meaning most people can't make the transition. It's akin to taking a beginning skier to a double black diamond run with the most advanced skiing equipment money can buy, pushing him off the cliff, and telling him to enjoy the ride. This would be a ridiculous way to learn a new skill, and it doesn't work for learning a new way of eating, either.

LEARNING NEW HABITS

I'm going to teach you new eating habits. As Mark Twain observed, "Habit is habit and not to be flung out of the window by any man, but coaxed down-

It's All About Control

. . .

If the real goal of a weight-loss diet is to *keep* weight off, most popular programs are dismal failures. Research has shown that from Atkins to Ornish, within one year of starting a new diet, 95 percent of all people either regain all the weight they've lost or even add a few extra pounds.[1–3] That's right; diets all succeed in the short term—and all fail in the long term because weight rebounds when people return to their old ways.[1,3,4] Not surprisingly, most popular diets employ similar principles and "tricks" to produce fairly rapid weight loss over a six- to twelve-week period. All involve taking control of your eating habits, whether calorie intake, fats, carbs, portion size, or even the times you eat. The list of things you have to control seems endless.

Did you ever wonder why almost all popular diet books say you can lose *x* amount of weight in six to eight weeks, but then don't tell you what to do next? And why, despite weeks of "good" behavior, you ultimately succumb to the lure of a triple cheese-bacon hamburger with a super-size order of fries or hot fudge sundae, and effectively say, "Screw the diet"? Yes, for about six weeks just about anyone can control any one—or even several—of these parameters, but eventually control always seems to be wrenched from you, doesn't it? Diets fail because they don't deal with who's really in control over the long term: your unseen autopilot.

stairs a step at a time." Research has shown that it takes at least six weeks of continuous practice to instill a new habit.[5] But unlike other programs with "one way" of doing things that also involve modifying your behavior for several weeks, each phase of Diet Evolution will guide you to new habits, to keep you on the straight and narrow as you restore your health. You'll learn how to change your habits almost imperceptibly, one step at a time. In doing so, you'll evolve new eating and behavior skills that allow you to survive in a new environment. I'll use the same techniques I use to turn young doctors into accomplished heart surgeons and will turn you into an accomplished expert on the care and feeding of the genetic program housed in your body.

Whenever you learn a new sport, you start simply and then advance according to your ability. As you move beyond beginner skills, you abandon the training wheels, props, or techniques essential to getting started so you can advance beyond beginner skills. So, too, in Diet Evolution, you'll eventually put aside the tools you'll initially use to *turn off* the killer genes and shut down the messages you once sent your genetic program. While these techniques will *stop* the destructive process and result in considerable and consistent weight loss, a very different follow-up technique is required to *reverse* the damage and *rebuild* your health. This second technique, if you wish, can be taken even further to enable you to *turn on* your genetic longevity program. Rest assured; you can always revisit the techniques acquired initially if and when you feel out of control, just as the snowplow is a great fallback technique even for accomplished skiers when all other attempts to stop fail.

MYTH BUSTING

Before you actually begin the program detailed in the following chapters, get ready for something really different. In Diet Evolution, you'll take an axe to some myths about healthy eating, which will further distinguish it from many other diets, particularly those focused on calories and low fat. These myths may seem like heresy at first, but in the coming chapters, you'll understand why much conventional advice about eating is off base. Consider these:

MYTH: Drink three glasses of milk or the equivalent each day.
REALITY: Most adults are not programmed to drink milk.

MYTH: Orange juice is a great source of vitamins.
REALITY: Drinking any fruit juice is like drinking liquid sugar.

MYTH: Whole grains, particularly oatmeal, are heart healthy.
REALITY: Eating ground-up grains is like consuming sugar.

MYTH: Three square meals is the way to go.
REALITY: Five smaller meals or three meals and two snacks help control hunger throughout the day.

MYTH: Because bananas contain potassium, you should eat them if you have heart problems.
REALITY: A ripe banana's starch instantly converts to sugar in your bloodstream, making it the ultimate heart-unhealthy fruit. (A green banana is a different story.)

Goodbye to Fibromyalgia

. . .

Lynn had seen what was happening to her husband, who was following Diet Evolution and wanted to join The Club. A former high-power executive, at 64, Lynn was disabled from fibromyalgia, severe arthritis, and asthma, and taking among other drugs, steroids and drugs to lower her blood pressure and cholesterol. Like so many unfortunate women with such complaints, Lynn had wandered from doctor to doctor, having test after test, trying every painkiller and antidepressant known to man—as well as some I had never heard of—to no avail. To make matters worse, her weight had skyrocketed to 251 pounds. Although Lynn was 6 feet tall, fat still constituted more than 51 percent of her weight. Her story and appearance screamed metabolic syndrome!

Lynn's lab work confirmed that she was insulin resistant, and her triglycerides were a whopping 182. As with most of my patients, her various doctors had told her that her cholesterol of 128 was spectacularly low because of the statin drugs they were giving her, but they hadn't bothered to mention that her HDL was 33, which is dangerously low. Moreover, her extremely low level of the fluffy type of HDL, which helps clear bad cholesterol, was minute, putting her in the highest risk category for heart disease and stroke. There was no time to waste!

Eight months after Lynn started Diet Evolution, she shed 52 pounds, breaking through her psychological barrier of 200 pounds. Her blood pressure is 120/80 and she is off her blood pressure medications. Her resting heart rate is 80, down from 98 at her first visit. She has a smile on her face and her movements are no longer labored. As she left my office, she looked over her shoulder at me and said, "I guess you have already figured it out, but I'll tell you anyway. My fibromyalgia is gone." She was right; I already knew.

MYTH: Tomato or vegetable juice provides a serving of vegetables in every glass.

REALITY: Tomato-vegetable juice is drowning in sugar.

MYTH: Low-cal or no-calorie sweeteners help you lose weight.

REALITY: All studies show that artificial or no-calorie sweeteners stimulate the release of insulin, which makes you store more fat!

Before you embark on what may be the most exciting project of your life, let me provide you with a brief summary of the framework upon which you will build.

- *Plants produce phytochemicals that are designed to punish, warn, trick, or otherwise confuse plant predators or to protect the plant from damage.*

- *Animal genes evolved in constant contact with plant phytochemicals.*

- *The presence or absence of phytochemicals within our cells prevents or produces cellular dysfunction.*

- *Plant leaves and seeds are "good" for you because they're "bad" for you.*

- *Modern processed food and animal products are "bad" for you because they're "good" for your genes.*

- *Modern foods, particularly ground-up grain products, rapidly enter your blood stream as sugar and activate the "Store Fat for Winter" computer program.*

- *Modern fats are primarily composed of omega-6s, which stimulate inflammation and depression.*

Part Two

DIET EVOLUTION

Chapter 4

THE DIET AT A GLANCE

Welcome to the Diet Evolution program. This chapter will provide a brief overview of the three phases of the program and lists of what I call "Friendly Foods," "Unfriendly Foods," and finally "Foods to Banish Initially," which you will avoid for a while and can later add back in moderation. You'll undoubtedly be referring to this chapter frequently until you are completely familiar with both the foods you'll be eating and those to which you'll bid farewell.

Before you start any weight-loss program, you should see your physician. Ideally, he or she can run some baseline tests against which to track your progress. In addition to taking your blood pressure and measuring your heart rate, I recommend the following tests:

- Fasting glucose level
- Hemoglobin A1C
- Fasting insulin level
- Fasting lipid panel (preferably with fractions of LDL and HDL, lipoprotein(a) or Lp(a), Apo B, Lipo-PLA2)
- Homocysteine
- Fibrinogen
- C-reactive protein (CRP)

Berkeley Heart Labs, in Alameda, California, and other national labs perform these tests. (Most insurers cover these costs on a three-month's basis, but you may want to check with your carrier.)

If you are taking medications to lower high blood pressure or high blood sugar, you may need to be monitored closely. Be aware that if you follow my program properly, your blood pressure will begin to normalize, meaning you may experience an occasional dizzy spell because your blood pressure medicine is dropping your blood pressure too low. Likewise, if you take diabetes pills or give yourself insulin shots, a natural drop in blood sugar levels may cause dizziness. In either case, talk to your doctor about lowering your dose and the dizziness will disappear.

THREE MESSAGES, THREE PHASES

Other diet books may include two or more phases, but generally the way of eating is similar from phase to phase. Instead, I'm going to ask you to transition through three distinct eating approaches—each with an emphasis on particular types and amounts of foods—as well as modify your exercise program as you progress. So distinct are the phases that initially you'll use one type of food to help achieve results, and then abandon it or dramatically reduce the quantity once it has outlived its usefulness. As we do this, I'll explain how doctors Atkins and Ornish can both be right, although not at the same phase of health restoration. So let's begin by understanding how you must change the messages you've been sending your genes these many years, so you can convince them that:

You Don't Have to Store Fat for Winter Phase 1, known as the Teardown phase, will accomplish this by significantly changing your eating habits to kick-start your weight loss. You'll do this by concentrating on protein foods such as meat, poultry, fish, shellfish, fresh cheeses, seitan, tempeh, soy products, and as many leafy greens and other vegetables as you like. You'll get to snack on nuts and seeds, but you'll renounce all foods made with sugar or refined grains and other processed foods. I recommend you follow this phase for at least six weeks. For those of you with significant weight-loss problems, you can safely stay on Teardown for up to a year.

You're Not a Threat to Future Generations You'll do this in Phase 2, the Restoration phase, by increasing your portions of vegetables and simultaneously decreasing your intake of animal protein. This shift along with adopting other behaviors will "trick" your genes into thinking they don't have to kill you off just yet because you're clearly useful to have around. You'll continue your nut and seed snacks, as well as eat berries and certain other fruits in moderation, which you will have resumed after two or more weeks in Phase 1. Likewise, you can, if you choose, resume eating extremely small portions of whole grains and legumes. And it's important to note that a whole grain is just that—whole. If you grind up a whole grain, it isn't whole anymore, is it? Again, I recommend you follow this phase for at least six weeks, preferably staying in it until your weight normalizes.

Your Staying Alive Ensures Your Genes' Future You'll do this in Phase 3, the Longevity phase, by turning on your longevity genetic program with "calorie optimization," meaning eating foods with the greatest micronutrient density but the least number of calories. This means that you'll rely heavily on vegetables, particularly salads and other raw vegetables, but avoid as much as possible those that are calorie dense, including whole grains and legumes. You will continue to reduce your portions of animal protein, or vegetarian alternatives, effectively treating them as side dishes. Not only will you actually obtain plenty of protein from vegetables, but by consuming large amounts of plant micronutrients, you'll also turn on genes that protect you in the long term, laying the groundwork for longevity, health, and vitality. Once you've achieved your goal weight and your health has improved, you'll learn how you can offset the occasional brief departure from the program without resetting the destructive genetic programs. This phase lasts the rest of your life.

The three phases of Diet Evolution are designed to teach you how to change your habits slowly and carefully, almost imperceptibly, one step at a time. While they have distinct objectives, the three phases work as a continuum, enabling you to gradually evolve your diet and behavior. Each chapter in Part Two focuses on techniques that will, in effect, retrace the actual evolution of the human diet so that you can re-evolve your own diet.

You can accomplish the first two phases in as little as ninety days—and move on to a permanent way of eating in Phase 3, unlike a diet you leave behind after getting the results you were looking for—or not. This is possible because new cells replace 90 percent of your existing cells every three months. Give your genes enough new materials with which to construct cellular components, and you can rapidly build a "new you." Trust me. Convincing genes that are only looking out for themselves that you hold the key to their survival will activate changes in all your programming that will restore your health and keep you thriving. This is not an idle promise, but one backed by ironclad research.[1]

THE FITNESS COMPONENT

Just as our eating habits have changed through the ages, so too have our patterns of movement or exercise. I've become convinced that our current "cardio"-focused exercise approach duplicates that of low-ranking primates, desperately struggling to make the grade, and the effort produces little in the way of results. In Diet Evolution, you'll learn how to evolve your exercise program to mimic what successful animals—including humans—do and have done for eons.

THE CASE FOR SUPPLEMENTS

I'll recommend specific supplements in the following chapters as well, but for now, I want you to understand why supplementing your diet is a key component of my program. As you've learned, plant compounds have tremendous impact on our cellular functioning, but today we consume fewer plant micronutrients than our ancestors did, making supplementing our diet with vitamins, minerals, and other micronutrients a form of health insurance. What about the government's recommended daily allowances (RDA) for certain vitamins and minerals? Don't I get everything I need in one bowl of Total cereal? RDAs were established after the introduction of refined white flour led to the discovery of vitamin and mineral deficiency diseases. But vitamin deficiency and vitamin adequacy are very different, just as living out an average life span and thriving for 80-plus years is not the same thing. My goal for you is not just a long life span, but also a long health span.

You may also have heard that recent studies do not show a clear benefit to supplementing with certain vitamins; but then why has the FDA recently upped the recommendations for a multi-vitamin, folic acid, and vitamin D for all Americans? To date, research has only scratched the surface regarding the usefulness of supplements.

There are two approaches to supplementation. The first is that if you eat plenty of fresh vegetables and fruits, they provide more of every compound than your genes can use. But consider this: foraging humans and our ape ancestors usually ate up to *200* different plants on a rotating basis. Today, the rare individual may actually consume twenty-five plants on a regular basis, but the average American eats about five, and that's counting French fries. Moreover, the vast majority of plants are grown with petrochemical fertilizers, which have existed for only the past fifty years. If you think for one

A doctor in his early forties on staff at the hospital where I practice, Thomas is thin and committed to his daily regimen of running. He ate rice, pasta, and whole wheat bread; drank skim milk and several glasses of pure vegetable and fruit juices a day; and eschewed all fat. Nonetheless, he was hypertensive, had an enlarging aorta, very low HDL ("good") cholesterol, and high triglycerides—despite being on statins and antihypertensives. He came to me after his own doctor had diagnosed diabetes. In our first meeting, he still couldn't get over the irony. "I'm doing everything right! How come everything is going wrong?"

I explained to Thomas that the lack of quality fats and greens in his diet, along with his daily jogging, were signaling his genes that he was an unsuccessful animal, working too hard for inferior food. His killer genes were activated. When I explained how my program would reverse the damage and restore his body to good health, Thomas was incredulous. What I was proposing was the opposite of everything modern medicine had taught him. But, what did he have to lose? He embarked on Diet Evolution. I met with Thomas after he'd run his own test of his cholesterol and blood sugar levels six weeks into the program. He said, "You're not going to believe this, but my blood sugar is normal, my triglycerides are normal, and my HDL is 54. It has never been above 30 in my life. It's just like you predicted!" He is a doubting Thomas no longer.

minute that sand-raised spinach has all the micronutrients of the greens raised in deep loam soil, I have oceanfront property in Palm Springs you might be interested in looking at as well. In any debate on supplementation, always remember that most plant compounds are still unnamed, and we don't fully understand how they interact with our genes.

The second viewpoint is just that: we literally have no idea what compounds, in what ratios, in what plants will combine to produce ultimate health and longevity, but we come closer each year to unraveling the riddle. Consider the example of the trace mineral selenium, which is present in nuts (particularly Brazil nuts) and yeast. The RDA of selenium was established by taking ten male college students, measuring the selenium content of their diet, and determining that they weren't ill. Unfortunately, the serum level of selenium in the blood of the average American is dramatically lower than that of the average French person, who incidentally has a much lower risk of diabetes and insulin resistance.

In each phase of Diet Evolution, I will suggest supplements that can enhance your health, work synergistically with your diet, and in certain cases, even help you lose weight. For more information on supplements, please visit www.drgundry.com.

WHAT TO EAT

Now, let's get to the meat of the matter, so to speak. In the chapters that follow, I'll explain in detail how much you can eat of "Friendly Foods" in different phases, when you can add back some foods initially restricted, and substitutes for those you will essentially banish from your daily diet.

FRIENDLY FOODS

PROTEIN

In Phase 1, your protein portions will initially be the size of your palm. As you progress, portions will become smaller and smaller.

Meat (preferably grass-fed)

- Beef filet, flank steak, stew meat, ground sirloin, round steak, jerky
- Lamb
- Pork tenderloin, ham, Canadian bacon, prosciutto (but no slab bacon)
- Wild game, venison, bison

Poultry (preferably free-range)

- Chicken
- Cornish game hen
- Duck
- Goose
- Turkey

- Turkey "bacon"
- Turkey and chicken cold cuts, preferably sliced and not processed parts
- Wild poultry

Fish (preferably wild, not farm raised)*

- Alaskan halibut
- Anchovies
- Freshwater bass
- Hawaiian fish, such as ono, mahimahi, opakapaka
- Mackerel
- Salmon, preferably Alaskan; also canned and smoked salmon
- Sardines
- Shellfish, including crab, lobster, squid, calamari, shrimp, scallops, clams, and mussels
- Trout
- Whitefish and perch (preferably from Lake Superior)
- Yellowtail and albacore tuna; also canned tuna

Dairy products (and substitutes)

Follow portion guidelines provided in Phase 1; you will diminish portion sizes as you proceed. Don't eat any processed cheeses or use milk, cream, or coffee lighteners. Aged cheeses should be eaten in extreme moderation (1 ounce or 1 slice a day).

FRESH CHEESES

- Farmer cheese (1 cup)
- Feta cheese (1/2 cup)
- Low-fat cottage cheese (1 cup)
- Mozzarella cheese, water packed (1/2 cup)
- Ricotta (1 cup)

*Note: For more information on all species of fish, including sustainability problems, the Monterey Bay Aquarium offers excellent regional guides to fish, noting which are endangered, at /www.mbayaq.org/cr/SeafoodWatch/web/sfw_regional.aspx.

AGED CHEESES

- Asiago
- Bleu cheese
- Cheddar
- Goat cheese (any kind)
- Gruyère
- Pecorino Romano
- Parmigiano-Reggiano
- Swiss

OTHER DAIRY

- Almond milk (plain or flavored, unsweetened only)
- Eggs (preferably omega-3, free-range preferred)
- Kefir
- Soymilk (plain or flavored, unsweetened only)
- Yogurt (plain, unsweetened only)

BIG FLAVOR, MINIMAL CALORIES

The best way to approach aged cheeses such as Cheddar or Swiss is with a vegetable peeler. Have you ever noticed how expensive restaurants shave Parmegiano-Reggiano cheese onto your salad? The taste is great, but the portion is small, so you save calories and they save money.

Soy and related protein sources

- Black soybeans
- Edamame
- Seitan (warning, this is wheat gluten)
- Tempeh
- Tofu (firm only)
- Tofu shirataki noodles

VEGETABLES

You can eat as much of the following as you wish.

Leafy greens

- Arugula
- Beet greens
- Chard
- Collards
- Dandelion
- Endive
- Escarole
- Kale
- Lettuce, including romaine, Boston, and all red and green lettuces
- Mustard greens
- Purslane
- Radicchio
- Spinach
- Turnip greens
- Watercress

Other vegetables

With the exception of pumpkin, you can eat as much as you want of these vegetables. Note that some should only be eaten raw and that tomatoes, avocado, and other fruits treated as vegetables are listed with fruits.

- Artichokes
- Bamboo shoots
- Beans (string, green, yellow)
- Bean sprouts
- Beets (raw only)
- Bok choy
- Broccoli and broccoli rabe

- Brussels sprouts
- Cabbage (napa, Chinese, green and red)
- Capers
- Carrots (raw only)
- Cauliflower
- Celery
- Chayote (vegetable pear)
- Chicory
- Cucumbers (including pickles)
- Fennel
- Garlic
- Ginger root
- Herbs (cilantro, basil, mint, sage, oregano, parsley, rosemary, thyme, etc.)
- Jícama
- Kohlrabi
- Mushrooms
- Olives
- Onions, shallots, scallions, chives
- Peppers (hot, such as jalapeño)
- Peppers (sweet bell, all colors)
- Pumpkin (up to 1 cup a day)
- Radishes (including daikon)
- Sugar snap peas, snow peas (but not shelled peas)
- Summer squash (patty pan, yellow crookneck, zucchini)
- Water chestnuts

GO FOR THE BURN

Eating jalapeños or other chiles, and hot, spicy foods in general, raises your metabolic rate, which enhances weight loss.

OILS

For dressing salads and other vegetables, my preference is extra-virgin olive oil, but many other oils add distinctive flavors. Some should be drizzled on food rather than used for cooking. Canola is useful for frying. Use what you need to flavor your foods. Do not use any prepared salad dressings, but if you must, use olive oil–based vinaigrettes or Caesar dressing. Avoid ranch dressing at all costs.

- Canola oil
- Flaxseed oil (do not heat)
- Grapeseed oil
- Hempseed oil (do not heat)
- Macadamia nut oil
- Olive oil (extra-virgin only, preferably unfiltered and cold pressed)
- Sesame oil (both plain and Asian roasted)
- Tahini (sesame seed paste), in small amounts only
- Walnut oil (do not heat)

CONDIMENTS

You can use the following condiments, but avoid any condiments such as chutney, ketchup, and barbeque sauce made with sugar, corn syrup, or any other natural or artificial sweeteners. Read the list of ingredients, and *when in doubt, leave it out.*

- A.1. Steak Sauce
- Fish sauce (*nam pla*)

- Lemon and lime juice
- Mustard (without honey or sugars)
- Salsa (preferably fresh)
- Salt and pepper
- Spices and herbs
- Tabasco and other hot sauces
- Tamari or soy sauce
- Tomato sauce or paste
- Vinegar (apple cider preferred, or any other type)
- Worcestershire sauce

NUTS AND SEEDS

Be sure you eat only raw, unsalted products, with the exception of peanuts, which should be roasted, and keep portions to 1/4 cup twice a day initially. I have omitted sunflower seeds because they contain too much omega-6 fat and cashews, which are high in sugars. Nut butters are very high in calories, so eat only occasionally and no more than a tablespoon twice a day. If you are allergic to nuts, soy "nuts" are an alternative.

- Almonds
- Brazil nuts
- Cocoa beans (cocoa nibs) or chocolate with greater than 70 percent cocoa
- Coconut, dried or fresh unsweetened only (very high in calories)
- Flaxseeds (ground)
- Hazelnuts
- Hemp seeds (or powder)
- Macadamias
- Peanut butter and other nut butters
- Peanuts (unless you're cooking with them, use only roasted peanuts)
- Pecans
- Pepitas
- Pine nuts (pignolia)
- Pistachios

- Pumpkin seeds
- Sesame seeds
- Walnuts

MEAL REPLACEMENT BARS AND POWDERS

These foods are great in a pinch, such as when you are running late for work, and enable many of my patients to stick with Diet Evolution, especially in the first phase. Still, they should be used only occasionally. Use brands that are both high in protein and low in carbs. Some taste very sweet, which can prompt an insulin response. Remember, you want to *retreat from sweet.*

- Most low-carb protein bars, including:
  - *Atkins Advantage*
  - *Doctor's Carbrite Diet*
  - *Pure Protein*
  - *Think Thin*
- Most low-carb protein shakes, such as Atkins Advantage and Pure Protein
- Hemp protein powder (unsweetened)
- Rice protein powder (unsweetened)
- Soy protein powder (unsweetened)
- Whey protein powder (unsweetened)

BEVERAGES

- Coffee
- Consommé, bouillon, clear broth
- Plain spirits without mixers
- Red wine
- Tea (black, green, white, and herbal)

UNFRIENDLY FOODS

VEGETABLES

These vegetables and others like them contain starches that turn to sugar quickly in your bloodstream. Some can be eaten raw, but not cooked, as indicated.

- Beets (cooked)
- Carrots (cooked)
- Corn (cooked)
- Peas (shelled)
- Root vegetables (parsnips, turnips, rutabagas, celery root)
- Sweet potatoes
- Winter squash (Hubbard, acorn, butternut, and others)
- Yams

"WHITE" FOODS

While not all are actually white, avoid these foods at all costs. *If it's "white," keep it out of sight.*

- Artificial sweeteners (Equal, Sweet 'n Low, Splenda, etc.)
- Candy (including sugar-free)
- Flour
- Frozen yogurt
- Ice cream
- Mayonnaise
- Milk (skim, or fat-free, are the worst)
- "No-added-sugar" foods
- Pasta
- Potatoes
- Ranch dressing
- Rice (including white basmati and most brown rice)
- Rice milk
- Saltines
- Soymilk, regular or "lite" (unsweetened is okay)
- Sugar
- White bread

"BEIGE" FOODS

These are also off the menu. *If it's "beige," better behave.*

- Bagels
- Blended coffee drinks
- Bread (including flat bread, pita, and whole wheat, whole-grain, and sprouted-grain products)
- Breaded food (any kind)
- Buns
- Cereals (hot and cold)
- Chips
- Cookies
- Crackers
- Deep-fried food (any kind)
- French fries
- Low-fat processed foods
- Muffins
- Pastry
- Pizza
- Pretzels
- Rolls
- Tortillas (flour or corn)

KILLER FRUITS

Avoid these sugar and calorie bombs.

- Dates
- Dried fruits of any kind (currants, prunes, blueberries, cranberries, etc.)
- Fruit leather/strips
- Mangos
- Pineapple
- Plantains

- Raisins
- Ripe bananas
- Ripe papayas
- Ripe pears
- Seedless grapes

OTHER FOODS TO AVOID

- Alcohol in a mixed drink, white or rosé wine, beer, malt liquors
- Fruit juice (all kinds)
- Honey, molasses, maple syrup, corn syrup, and other sweeteners
- Jam, jellies, preserves, condiments made with sugar
- Jell-O (including sugar-free Jell-O)
- Soft drinks, including sugar-free and diet brands
- Vegetable juice (all kinds)

FOODS TO BANISH INITIALLY

"BROWN" FOODS

These foods unquestionably slow weight loss. *If it's "brown," slow down.*

- Amaranth
- Barley
- Brown basmati rice (from India)
- Buckwheat (kasha)
- Bulgur
- Corn kernels (fresh and raw only)
- Farro
- Legumes (beans such as lentils, lima, garbanzo, navy, pinto, kidney, etc.)
- Millet
- Oats (whole or steel-cut, but not old-fashioned or quick oats)
- Quinoa

- Rye
- Soy/spelt/garbanzo bean flour (for coating only)
- Spelt
- Whole wheat berries
- Wild rice

Later in Phase 2, you will learn why it will be wise to think of cooked meats as brown foods as well.

FRIENDLY FRUITS

Serving size is indicated in parentheses. These fruits can be added back after the first two weeks in Phase 1, but their use will slow weight loss. Have no more than two servings of these fruits a day after reintroducing them.

- Apple (1 medium)
- Apricots (4 fresh)
- Avocado (1/2) (Yes, an avocado is a fruit)
- Banana—green-tipped only (1)
- Blackberries (1 cup)
- Blueberries (1 cup)
- Boysenberries (1 cup)
- Cherries (1 cup, about 10)
- Citrus—oranges, grapefruit, tangerines, tangelos (one whole, or half a grapefruit)
- Cranberries (1 cup)
- Currants (1 cup fresh)
- Grapes—but not seedless ones (1 cup)
- Guava—no guava juice (3 small)
- Huckleberries (1 cup)
- Kiwi—try it with the fuzzy peel on (1)
- Kumquats (2)
- Lychees (1 cup, about 5)
- Mulberries (1 cup)
- Nectarine (1)

- Papaya—green only, in salads (1)

- Passion fruit (1)

- Peach (1 medium)

- Pear—firm, not ripe (1 medium)

- Plums (2 small or 1 medium)

- Pomegranate (1/2 cup seeds)

- Raspberries (1 cup)

- Strawberries (1 cup, about 6)

- Tomato (1 medium)

Restaurant Rules

. . .

Don't take the easy route of thinking it's impossible to be a Diet Evolutionist when dining out. Just remember this saying taught to me by my good friend Tom Guy: *The menu just tells you what foods the chef has in the back.*

To keep you on the straight and narrow (as in your hips):

- Never let the waitstaff put bread on the table—you'll eat it!

- "Graze" on appetizers and salads, instead of an entrée.

- The meat or fish entrée is usually served over or with a starch and a vegetable. Ask that they hold the potatoes, rice, or pasta; instead, ask for a double portion of veggies. Scan the menu; if a "Friendly Vegetable" is served with a different entrée, just ask for it to be served with your choice instead.

- Order vegetable side dishes. Don't fret if the spinach comes "creamed"; it's the baked potato that will do you in!

- For dessert, order berries or skip it altogether. If that won't do the trick, order one dessert for the table and split it four ways.

If the chef won't help you restore your health, you probably shouldn't go back!

Chapter 5

THE FIRST
TWO WEEKS

NOW that you understand the objectives of the three phases, I know you're eager to get down to the real business at hand. What can you eat and what should you avoid in the crucial first two weeks? You'll get most of your calories from protein (meat, fish, poultry, dairy, and other sources listed as "Friendly Foods" in the previous chapter). You'll have a portion of protein about the size of the palm of your hand for breakfast, lunch, and dinner. Don't worry if you're a vegetarian or vegan; there's plenty for you to eat, too. Nor need you worry too much about how you cook (or don't cook your food) at this point. Preferably, sauté your protein in olive or canola oil or grill it. Preferred vegetable cooking methods are steaming, sautéing, or stir-frying in a wok. Please, just don't deep-fry any food or overcook any vegetable.

You have probably failed at other diets, just as I have, not because you're weak-willed or have a problematic or sluggish metabolism, an underactive thyroid, a hormonal imbalance, or whatever, but because the initial teardown phase of the diet did not evolve into a restoration phase. Or, conversely, you tried a useful rebuilding diet as a first phase, but without a well-prepared foundation, there was no structure upon which to build.

I'm going to ask you to initially make specific food choices that will "destroy" the old messages you've been sending your genes. During this phase, you'll be deactivating the destructive "Winter Is Coming" message that triggers the "Store Sugar as Fat" genetic program. Instead, you'll inform your genes that "Winter Is Now," (WIN), which activates the burning of stored fat. This second message initiates the demolition process and allows you to slim down. As you'll soon see, whenever you activate the WIN program, you'll win your battle with excess weight.

I call the first six weeks of my program the Teardown phase because the process of restoring your body to optimal health has much in common with that of restoring a house to prime condition. The first two weeks are the most intensive. When you redo an old house, you knock down unwanted walls, remove corroded wiring and plumbing, expose the studs, and often shore up the foundation. Even though this is usually the most backbreaking, hazardous part of the job, omit these foundational efforts and the end result will be disappointing, superficial, and impermanent. Anyone who has hastily painted over an unprepared surface, only to be greeted by déjà vu all over again a few months later, has learned this lesson the hard way. Sure, it was easier to slap on a new coat without sanding and prepping, but the effort saved in the short term didn't pay off in the long term. It's okay to admit you've done this. I've done it myself too many times. More important, I did it to myself all too many times.

I could just as easily have called Phase 1 the Pruning phase. In the winter, you may cut away up to 75 percent of the active leaf-bearing surface to keep fruit trees strong and producing fruit. Ouch! But like clockwork, come spring, out pop hundreds of new shoots, and by summer they seem more vigorous than ever. Pruning is essential, but continually pruning is counterproductive, just as removing all the supporting walls in a house will cause it to collapse. This failure to realize the need for two very different techniques at different times explains why most well-meaning diets ultimately don't work.

THE POWER OF PROTEIN

To convince your genes that "Winter Is Now," in the first two weeks of Diet Evolution, you'll focus on eating protein and vegetables, which as you can see from the lists of acceptable foods in the previous chapter offer you a remarkable array of choices. In addition to the palm-size portions of protein you'll eat at each meal—yes, big-handed guys get more than small-handed gals—you'll be eating plenty of leafy green vegetables and two snacks of seeds or nuts. And in place of these palm-size portions of meat, poultry, fish, and meat substitutes, you can also have the following servings of protein:

- 1 cup of fresh cheese, such as cottage or ricotta
- 1 ounce of aged cheeses, such as Cheddar or Swiss
- 1 cup of plain unsweetened yogurt
- 1 cup of plain or flavored unsweetened soymilk
- 1 cup of plain or flavored unsweetened almond milk
- 2 or 3 eggs (preferably omega-3 and free-range), up to 4 per day
- 1/2 cup black soybeans (not to be confused with black beans) or edamame (green soybeans)
- 1 package of shirataki tofu noodles

Turn to page 174 for daily meal plans for the first two weeks of Phase 1. In Part Three you'll also find recipes suitable for the various phases, as well as helpful cooking and preparation tips.

You'll start off with a high-protein diet for two reasons. First, animal protein would have contributed a greater percentage of calories to your ancestors' diet during the winter or dry season, when plants were usually dormant. Probably those animals were leaner than today, but let's not get into a debate about whether Stone Age animals were fat or thin; in the Teardown phase, it doesn't matter. Having said that, remember that well-marbled meat would have been available only during late summer and fall, when the animals that humans hunted had gorged on fruits and grains, activating their own "Store Fat for Winter" program.

Eating protein, on the other hand, signals your genes that "Winter Is Now," which in turn activates your "burn fat" program. That's why most of the cuts of meat and poultry I recommend are relatively low in fat. Protein has another distinct benefit in the Teardown phase and is the secret behind initial rapid weight loss on the Atkins, South Beach, and other high-protein diets. Compared to fat and carbohydrate, protein is the *least* efficient source of energy, requiring a tremendous amount of digestive work and breakdown of complex protein molecules into amino acids. In fact, Arctic explorers in the nineteenth century who had to resort to eating primarily lean rabbits and rodents actually starved to death, a phenomenon termed "rabbit starvation" or "protein starvation."[1] Why is protein such a lousy calorie source? The process of breaking it down into usable fuel actually consumes 30 percent of the calories, primarily as heat production, making them unavailable to you. Fortunately, we can use this trick to get more bang for our buck in initial weight loss.

This heat production has an additional benefit: it results in early satiety—the feeling of fullness—compared to diets higher in carbohydrate or fat. Indeed, when volunteers are put on a high-protein, high-fat, or high-carbohydrate diet—all of which are equally pleasurable and palatable—those on the high-protein program always consume the fewest calories.[2] When your diet centers on protein, not only will you eat fewer calories, but you'll also get the double bonus of fewer *usable* calories. Let's do the math. If you consumed 2,000 calories of pure protein daily, you'd have only 1,400 calories of usable energy. Again, the secret of high-protein diets is that you can eat more protein than fat or carbohydrate and get "free" calories. That's why I like to say, *you get to cheat if you eat protein or meat.* Don't believe those nutritionists who claim a calorie is a calorie. When it comes to protein, it's simply not true.

THE SECRET OF QUICK WEIGHT LOSS

Yet another benefit of eating protein during the Teardown phase is that proteins can be broken down into sugar, through a process called gluconeogenesis. Here's how it works: the primary fuel your brain uses is glucose, or sugar—no surprise, there—which it prefers to all others. Your brain will switch over to burning fat, as anyone who has followed a true high-fat/low-carb diet knows, but it's not happy about it. That's why most high-fat/high-protein dieters get a headachy feeling for a few days in the initial phase as the brain acclimates to burning fat for energy. You may get the same feeling in the first few days of the Teardown phase.

But here's what you didn't know: since your brain is never happy with this circumstance, it's constantly transmitting chemical distress signals asking for glucose. Your body responds initially by liberating your residual sugar stores, known as glycogen, from your muscles and liver. (Using up this limited supply of glycogen is what causes marathon runners to "bonk" at mile 20.) Each molecule of glycogen in these tissues is bonded to a water molecule, so as the glycogen is liberated, the water rushes into your bloodstream and passes through your kidneys. Bingo! Now you know why you get a sudden "miracle" weight loss when you start most diets. Believe it or not, you've stored this water because in ancient Africa, winter was a dry season; you're designed to store fat and water when your genes perceive winter is coming. That bloated feeling and your high blood pressure usually disappear when you start Phase 1, for this reason.

Thanks to the liberation of glycogen, you will experience a 3- to 5-pound weight loss when you start Diet Evolution. This occurrence is so reliable that I use it to track whether my patients are correctly following the Teardown phase. If they lose this weight, I can tell they're on track; if not, I'll know some-

Father-Daughter Teamwork

. . .

I first met Melissa in the hospital waiting room after operating on her 88-year-old grandmother for severe coronary artery disease and a narrowed aortic valve. Grandma was going to be fine, but I didn't like what I saw with Melissa. As a 20-year-old college student, she was 5' 2" tall and weighed more than 300 pounds. Her father, Robert, was also closing in on 300 pounds. Each clenched a large, low-fat Frappuccino. While I assured them that their relative was going to be fine, I also told them they had been given a great gift that day, because they could foresee their future if they didn't change their eating habits. I gave them my card.

Thankfully, father and daughter heeded my warning and started Diet Evolution together. Both dutifully checked in with me once a month. After eight months, Robert is down 72 pounds and continues to lose weight—a little faster than I would prefer, but he does have a long way to go. His high blood pressure and daily headaches are history, and he has thrown away his diabetes and blood pressure pills. Now that he no longer has erectile dysfunction, he also has a new love life.

After an initial 30 pounds of easy weight loss, Melissa hit her first plateau, which coincided with spending the summer at home. While her dad continued to lose 8 pounds a month, her weight just hovered. I assured her that summer is the most difficult time of the year to lose weight, and advised her to settle in to her new weight and hold it steady for the summer. And so she did. When she returned to the office in the fall after being back at school for a month, she was 9 pounds thinner than her pre-summer weight. It was happening just as I had assured her. Melissa has a new boyfriend and there's talk of a future together after graduation. If she keeps using the Diet Evolution teachings, that future will include good health.

---

thing is amiss. In the first two weeks, patients shed pounds quickly, but their percentage of body fat on an electronic body-fat monitoring scale actually goes up. How can you lose weight but see your body fat rise? Don't worry. It's just a trick. As glycogen is burned for energy and water is flushed from your tissues, the machine assumes that there is a greater percentage of fat.

Remember, your brain is constantly looking for glucose. After tapping the glycogen stored in your muscles and liver, where does the brain look next? Unfortunately, it sends signals to your muscles, which are composed of protein, to break down and, via gluconeogenesis, make sugar. In fact, studies of Atkins dieters show that up to 50 percent of the weight loss in the first six months of these diets is muscle.[3] Yikes! After all you've learned about the importance of muscle mass and what it tells our genetic computer program about whether or not you should be kept around, that's the last thing you need. Moreover, since muscles burn calories, if you lose muscle while trimming pounds, no wonder you regain them so rapidly after you stop dieting.

This is a classic example of a good idea, using a high-protein/high-fat program to shut down the self-destruct genes and fat-storage system; but that diet, taken too far, damages the foundation that we wish to preserve. In Phase 2, the Restoration phase, you'll learn how to preserve and even build your muscle mass as you continue to lose fat. Building muscle has an additional benefit: it informs your genetic programming that you're out there hauling buffalo or roots (or grass-fed beef and buffalo mozzarella Caprese salad from Whole Foods) back to camp after wrestling a saber-toothed tiger (or that 300-pound NFL linebacker) for your share of the kill. In other words, you're useful to keep around.

THE NATURAL CHOICE

I recommend free-range poultry and grass-fed beef, as their food sources are closest to those found before the advent of modern farming methods, which rely primarily on grain. Grazing animals are likeliest to have the most beneficial ratio of omega-3s to omega-6s, as well as being a great source of plant phytonutrients in their tissues. Likewise, you're better off eating wild-caught than farm-raised fish.

BECOME A LEAN, GREEN EATING MACHINE

I promised you that you would be able to eat plenty of food and never feel hungry on Diet Evolution. While protein will provide the bulk of the calories in the Teardown phase, vegetables will supply most of the volume and micronutrients. No need to measure your veggies—you can eat "Friendly Vegetables" (see page 61) to your heart's—or should I say, your tummy's—content. If you cook the veggies, the portion will look deceptively small. The rule of thumb is that 3 to 4 cups of raw leafy greens equal 1 cup of cooked greens. Do make sure to eat some of your greens as salad. I'd like to see you eat *a plateful*— and make that a dinner plate, not a dinky little salad plate—of leafy green veggies with lunch and another one with dinner. You can place your protein portion on top of the vegetables. Have vegetables at breakfast, too, if you wish. By the end of the day, you should have consumed the equivalent of one 5-ounce bag of dark leafy greens. You can also consume as many other "friendly vegetables" as you want, but unless you're already a big veggie eater, you'll feel better if you increase your consumption gradually to let your system adjust to this new way of eating. Here are two other important instructions:

- You can and should have a mid-morning and mid-afternoon snack of a handful of nuts and/or seeds. Why nuts? Read on and you'll find out. If you're allergic to nuts, you can likely substitute soy nuts.
- Drink 8 to 10 glasses of water daily.

Finally, many of you will be pleased to know that women can have one glass of red wine a day and men can have two. Or if you prefer, women can imbibe one shot glass of spirits—and men, two—minus mixers. In other words, tequila, *sí*; Margarita, *no*.

VARY YOUR VEGGIES

Getting our nutritional needs met with a variety of foods is a protective mechanism that is also hardwired into our genetic code. It probably explains why single-food diets, whether the Grapefruit Diet, the Egg Diet, the Cabbage Diet, the Foie Gras Diet (yes, there is one), work well initially. You inherently eat less and less of a single food, if that's all that's available, to protect against the possibility of overloading your liver's detoxification system.[4] Your genetic programming "makes" you eat progressively less and less of the same food to protect the body your genes are occupying, just as it made you take a breath when I asked you to hold it in Chapter 1.

DON'T CONFUSE HIGH PROTEIN AND HIGH FAT

In the first two weeks of the Teardown phase, eating lots of protein—whether animal or plant based—will ensure you lose weight. You don't have to be afraid of how "unhealthy" it might be. Remember, we're pruning the tree, demolishing the old, tired parts of the house. Consider protein as the crutches and training wheels that will get you started on a new lifestyle; use it now and we'll worry about backing off later. But understand that eating a diet high in protein is not the same as one also high in fat. There is no evidence that a high-fat diet is useful in the long term; moreover, the wrong kinds of fat can be deadly.

A NUTTY GIFT

When your brain goes looking for glycogen and you don't want it to raid the stores in your muscles, what do you do? Give your brain small bits of glucose and protein in the form of nuts or seeds so it will leave your muscles alone. Nuts or seeds will also help you deal with cravings, so you can stay satisfied and motivated.

About 1/4 cup of raw walnuts or other nuts or pumpkin seeds—that's one handful—as a mid-morning snack and another handful as a mid-afternoon snack contains sufficient protein and glucose to keep your brain happy without depleting your muscles. You can substitute nut butters occasionally—just make sure they have no added oils—but remember that they're processed foods and didn't exist a century ago. Moreover, whole nuts and seeds have the added bonus of *not* being completely digestible, meaning you absorb only about two-thirds of the calories. My patients are shocked by the remarkable hunger-erasing effect of raw nuts and seeds, which blunt the pre-meal rise of ghrelin, the chief hunger hormone.

Don't use roasted or salted nuts or seeds, with the exception of peanuts. (Raw peanuts can contain aflatoxin, a dangerous mold.) Because the omega fats that nuts contain oxidize rapidly, processors roast nuts to keep them "fresh," but doing so destroys any antioxidant benefits. And salt, as anyone who has had a close encounter with a bowl of salted cashews will attest, makes you want another handful, then another. But as a starter, do try Dr. G.'s World-Famous Nut Mix (page 196). As your diet progresses, the increasingly large amounts of green food you'll be consuming will also *slowly* break down into glucose during the day. You'll learn this green-foods trick in the weeks to come.

Road Warrior Survival Tools

. . .

Many of my patients approach their first business or recreational trip with trepidation; how will they find Diet Evolution foods on the road? My primary rule still applies: Do what you can, with what you've got, wherever you are. The biggest challenges are often hotel breakfasts or eating at airports. My advice is to leave home with high-protein, low-carb bars in your suitcase. Grab a coffee or tea, maybe an apple or orange (after the first 2 weeks), and enjoy your bar. Here are some other tips:

Breakfast: Many hotels have those awful scrambled eggs on the buffet bar. Don't sweat it! Eat them, and leave the hash browns and toast alone. Or order poached eggs or an omelet with veggies. Substitute sliced tomatoes for toast and potatoes, and make sure they don't serve you the complimentary orange juice. If they pour it, you'll drink it.

Airports: Almost all airports now have some sort of chain dining establishment, whether it's TGI Fridays or Chili's, or a local version of such. Order a Caesar salad with no croutons. In a pinch, order buffalo wings, or strips, and a salad. Don't delude yourself that it's a great meal for you, but it won't trigger the "Store Fat for Winter" genetic program.

Chinese, Mexican, or other fast-food joints: Get what you want, but no rice, beans, bread, or taco chips. I've even had it "my way" at burger shops: Believe me, they have lettuce leaves in the back that they will wrap around your burger.

Snacks: I always pack a generous supply of Dr. G.'s World-Famous Nut Mix (see page 196 for the recipe): some in my briefcase, the rest in my suitcase. If you forget, almost every newsstand/snack shop in airports or on the road has prepackaged nuts. Grab the raw almonds, and get your hands off the bigger and cheaper bags of peanuts or sunflower seeds; stay away from the cashews and the "healthy" trail mix. Advantage Bars, or another high-protein, low-carb bar, are almost always available next to the candy.

Coffee bars: Instead of that latte or cappuccino, order a "double short cappuccino," which is actually an espresso macchiato—an espresso with a dash of foam. You'll also save money; it's cheaper than either a latte or cappuccino and better for you.

KILL SUGAR CRAVINGS: SUPPLEMENT YOUR DIET

In the first two weeks, three micronutrients—selenium, cinnamon, and chromium—can help you resist the urge to eat sweet things, thereby helping diminish insulin resistance. Taken in combination, they kick-start any weight-loss program and have been shown to dramatically reduce both glucose and insulin levels in all the patients I've studied.

Selenium Higher selenium levels are associated with reduced insulin resistance. In fact, supplementing with this trace mineral has improved my own and my insulin resistant patients' glucose levels.[5]

 Typical daily dose: 200-400 mcg

Cinnamon In the past five years, solid research, first in India and now elsewhere, has demonstrated the value of the ground-up bark of the cinnamon tree in dramatically lowering glucose levels of diabetics and insulin-resistant individuals.[6,7] The spice acts just like insulin on insulin receptors, allowing cells to take up sugar.

 Typical daily dose: Start with 500-1,000 mg (1/4-1/2 teaspoon) twice a day

Chromium The trace mineral chromium interacts with insulin receptors and improves insulin's action.[8] Moreover, research has shown that rats on chromium-deficient diets die young, while rats given chromium supplements have longer than normal lifespans.[9]

 Typical daily dose: 400-1,000 mcg of chromium picolinate or GTF chromium

---

SPICE AS NICE

Try mixing ground cinnamon with cottage cheese or yogurt, or stir your coffee with a cinnamon stick—just leave it in for at least 5 minutes. And no, getting your cinnamon in the form of a Cinnabon doesn't count!

BENEFICIAL BACTERIA: SUPPLEMENT YOUR DIET

Since you're starting to restore your body, you'd better start reinvigorating a huge (literally) contributor to the proper functioning of your body—the pounds and pounds of bacteria in your colon. No, it's not a cesspool that needs to be periodically cleaned out. But if you have ever taken antibiotics or eaten a heavily "white" and "beige" food diet (see pages 66 and 67), you've filled your colon with bacteria and fungi that are sending abnormal chemical messages into your bloodstream 24 hours a day. Really. To repopulate with beneficial bacteria, avoid those foods and take a probiotic that includes several of the normal bacteria such as *Lactobacillus acidophilus, L. rhamnosus, L. salvarius*, and *Bifidobacterium bifidum*, in a growth-promoting medium like maltodextrin.

Typical daily dose: Take capsules with at least 4 billion bacteria once a day for the first month of Diet Evolution, and whenever you are traveling or during and after you take antibiotics.

Equally important, these healthy bacteria need fertilizer in the form of prebiotics. The most beneficial are fructo-oligosaacharides, better known as FOS. Need help remembering their name? Just think of them as Friends of Steve (FOS); they'll soon be friends of yours! Some of the most common are inulin, derived from chicory, and yacón, an Andean root vegetable that is a cousin of the sunflower. Available powdered or as a syrup, yacón has the remarkable property of being sweet without being absorbed by the body as a sugar is. Other simple sugars that are not FOSs, but still are useful, are d-ribose, d-mannose, and xylitol. "Friendly Vegetables" that are high in FOS include garlic, onions, leeks, mushrooms, asparagus, and artichokes. FOSs will also help you absorb more magnesium and calcium, improve your immune system, and lower your cholesterol.

Typical daily dose: 500–1,000 mg twice a day

In this chapter, I've focused on all the delicious foods you can eat, many of them in unlimited quantities. In the next pages, you'll come to understand exactly what foods you need to avoid to pare away pounds and begin the process of restoring your health and vitality.

Chapter 6

WHAT'S OFF THE MENU?

In Phase 1, you'll say good-bye to most sugary and other high-carbohydrate foods that provide calories but virtually no micronutrients. Instead, you'll be eating a host of nutrient- and fiber-rich micronutrients in the form of "Friendly Vegetables." The foods you'll be eating and those you'll be giving the boot in Phase 1 take you back to the diet that prevailed roughly a century ago, before commercially milled grains and other processed foods began to change the way humans ate. As you move through the phases of Diet Evolution, you'll continue to evolve your diet backward in time until it emulates the way our early ancestors ate and thrived. Just in case you are unsure about which foods are off limits, review the "Unfriendly Foods" lists on page 65. They include:

- All foods I've categorized as "beige" or "white," including pasta, rice, potatoes, milk, ice cream, crackers, chips, all baked goods, cereals, and candy
- All foods containing sugar in any form
- All soft drinks (including diet, low-cal, lite, and sugar-free) and alcoholic mixed drinks
- All fruit and vegetable juices
- White wine and beer, which have residual sugars

Don't groan! Don't tell me how much you love these foods. We all love them. As you learned in Part One, we—and any society or animal exposed to them—are programmed to do so. These foods feed the most powerful genetic program you run. It's no wonder that eliminating them seems daunting. But, once again, we're going to trick your genes into enjoying the ride. In the first two weeks, you'll break your (actually your genes') addiction to such foods. Get used to it, because these foods are never coming back, except as minor exceptions once you have your weight and health completely under control. When that time comes, as so many of my Club members and I have discovered, you may no longer be tempted by these foods.

Also, for the first two weeks, eliminate:

- All fruits, including berries
- What I call "brown" foods, such as whole grains and legumes such as beans and lentils
- Certain so-called vegetables that are really fruits, such as tomatoes, avocado, and eggplant
- Cooked root vegetables like beets, carrots, and celery root

WHY GIVE "WHITE" AND "BEIGE" FOODS THE HEAVE HO?

If the "white" and "beige" food lists—as well as the "brown" foods—look suspiciously like a "low-carb" reprise, you're right on target. Not to worry. I'm not asking you to count carbs, control carbs, or worry about their glycemic load. Nor need you limit your intake of leafy green and other vegetables, as you do on most low-carb diets. All you need to remember is that the purpose of the Teardown phase is to convince your genetic programming that winter is *not* coming (it's here) and therefore your genes don't need to send out the "store fat" signal. Eating sugary and other high-carb foods convinces your genes that you've just encountered a fruit tree laden with sugary fruit, and as you've learned, that's the last message you want to transmit. Instead, eliminating these foods as well as fruit sends them the "Winter Is Now" signal, telling them to burn fat rather than store it.

You keen-eyed observers will notice that there are a few exceptions to the "white," "beige," and "brown" rules in the food lists, but they're just that— exceptions. In the first two weeks, the idea is to avoid all foods that contain sugar or starches in any form, along with any sweet-tasting foods or ones that rapidly convert into sugar in your bloodstream. Some of these foods may not literally be white or beige (like a chocolate biscuit bar), but their essential ingredients are white sugar, flour, or both. Most beige foods are just baked sugar or flour, or are made to appear "healthy" by the addition of such ingredients as whole wheat flour. Don't be deceived. In general, if a package trumpets the words "all natural," "fat free," "no cholesterol," "old fashioned," "heart healthy," "sugar free," "no added sugar," or the like, run the other way. Clearly, I cannot list every food you may come across that has no place in your diet, so, *when in doubt, leave it out.*

It's important to understand that the foods you must avoid completely have nothing to do with whether they're "good" or "bad" for you; at this point, I don't care. My new patients love to tell me how they eat only organic and/or whole-grain bread, bagels, and pasta, so how could their blood pressure be elevated and their cholesterol be so high? I'm here to tell you that regardless of where you bought them or how the grains were grown and milled, these foods activate the "Store Fat for Winter" program.

GIVE FRUIT THE BOOT AND RETREAT FROM SWEET!

Now, for the final death blow to your "store fat" genetic program and the busting of a major myth: You're going to omit fruit for at least the first two weeks, and longer if weight loss is your primary goal. That's right, all of it. I don't care that fruit is loaded with micronutrients and fiber, or that it's considered healthful. Avoiding fruit convinces your body to deactivate the "Store Fat for Winter" program, which is all I'm concerned about at this time. Almost every new "health-conscious" patient I see with high blood pressure, high cholesterol, and a bulging tummy loves fruit. Just remember that doing the same thing over again and expecting a different result is insane. This time around you're going to do things differently.

Juices are particularly lethal because they activate your "Store Fat for Winter" program. You've been assured for years that fruit juice is a convenient way of getting those pesky five servings of fruits and vegetables each day, but, sorry to say, you've been duped. Ask any diabetic who has to take insulin shots what she should do when she has an attack of low blood sugar. That's right, drink OJ. It, or any fruit juice for that matter, will raise your blood sugar so fast and so high that I couldn't put a tourniquet around your arm, find a vein, and inject 50 percent dextrose into you any faster. Once again, if your blood

. . .

Fran is one of my health-nut patients. A slim 64-year-old, she worked out regularly and had the muscles to prove it. She was a vegetarian who believed that humans are designed to eat primarily fruit, and did she ever, consuming five to ten servings a day. But she had two problems: a very high cholesterol profile and a slow-growing type of bone cancer. Her total cholesterol was 277 with an LDL ("bad") cholesterol of 191 and an HDL ("good") cholesterol of 59. I told her that she hadn't inherited her high cholesterol; instead, her heavy fruit diet was activating the "Store Fat for Winter" program. Yes, her vigorous exercise program was keeping her buff, but the underlying problem was still there.

Fran didn't believe me. I asked her to trust me and eliminate fruit from her diet for a two-week period. She wouldn't agree, but did promise to drastically curtail her fruit consumption. Two weeks later, we repeated her blood test. To her shock, her triglycerides had dropped 40 points to 76, while her total cholesterol fell 56 points to 221; her LDL also fell 50 points to 141, while, sure enough, her HDL, the recycling cholesterol, rose 7 points to 66. In two weeks. Like everyone else who has followed Diet Evolution, when she changed the incoming message to her genes, telling them, "winter is here, burn fat," she experienced a dramatic drop in her cholesterol. Fran's cancer indicators had been rising, but when I saw her six months after starting Diet Evolution, her counts were the lowest they had been in years. So am I also saying that all that sugar was feeding her cancer? It sure looks that way.

sugar rises quickly, what does your body assume has happened? Right: You've just found the ultimate fruit tree; winter is coming; time to store fat. Next time you take a trip to the zoo, look in the cages. See a juicer? How about a blender for smoothies? Most animals eat whole fruit, but no juice.

If you're habituated to sweet tastes, particularly from artificial sweeteners, I don't want to shock your system with an abrupt change. Instead, I have a trick to lessen the hold these sweeteners have on you, just as they did on me.

GOT MILK, NOT

Milk is out as well. But doesn't milk "do a body good"? Not past infancy, it doesn't. All mammals are designed to develop intolerance to milk after infancy, thanks to a gene that switches off and stops manufacturing lactase, an enzyme that digests lactose, the sugar that makes milk appealingly sweet to babies. Without lactase, milk remains undigested, causing cramps and diarrhea—the classic signals of lactose intolerance. An adolescent mammal that suddenly gets cramps when dining at Mom's milk bar stops drinking very quickly. Roughly one-third of the world's population is *not* lactose intolerant because about 3,000 years ago a Siberian yak herder was born with a simple mutation in the gene that normally shuts down lactase after infancy. This guy could continue to get nourishment from yak milk without the unpleasant side effects, allowing him to store more fat for the winter, so he survived but his lactose-intolerant buddies may not have been so lucky. His descendants kept this advantage. If you can drink milk without adverse effects, you're actually a mutant and your lactose-intolerant friends carry the "normal" gene.

There's another, hidden downside to milk. It contains a hormone called insulin-like growth hormone (IGF), designed to stimulate cells to grow. All mammals secrete IGF into their milk to make their offspring grow quickly. This is great for an infant, but continue to consume IGF as you age and certain cells are continuously stimulated to grow—namely, breast gland cells, prostate gland cells, colon cells, skin cells, and cells lining your blood vessels and joints. Is it any wonder that milk consumption in adulthood is correlated with increased rates of breast, prostate, and colon cancer? Increased incidence of atherosclerosis is another association: Remember those breast-fed infants with plaques? That's right, milk. Don't be misled by labels that say words to the effect "Our cows are not treated with bovine growth hormone (or BGH)."

All cows' milk contains BGH (another name for IGF) to make calves grow, regardless of whether or not the mothers were injected with that hormone.

What else does IGF do? It acts just like insulin in your liver and turns on the "Store Fat for Winter" program. And guess what, skim milk has even more sugar than whole milk, and that sugar promotes the release of even more insulin, which—as you now know—commands your liver to make fat. The exception to dairy products is cheese, particularly fresh cheeses. It appears that cheese does not contain IGF; and certain fresh cheeses, like ricotta, farmer cheese, cottage cheese, and fresh mozzarella (the soft stuff, not the cheese sticks or the kind you slice on pizza) are a great source of protein, which is why you'll find them in the "Friendly Foods" list. You can eat aged cheeses in extreme moderation, meaning no more than 1 ounce a day—that's a single slice or a 1-inch square cube.

DEATH BY FRUCTOSE

Before you pop the top on your favorite soft drink, read the list of ingredients. Odds are it's sweetened with high fructose corn syrup. Unlike whole fruit, the drink is devoid of fiber so nothing slows the absorption of fructose in your stomach. Because you don't feel full, you continue to sip away—no wonder people actually spring for 64-ounce drinks. All the while, your liver churns out triglycerides to deposit in your fat cells and artery walls. Your genes program you to seek out such beverages, but they will kill you and you're literally drinking it up!

THE HIDDEN DOWNSIDE OF "BEIGE" AND "WHITE" FOODS

Remember, you're using a proven program that has worked for people regardless of their age. I can tell you that I have seen dramatic results in just six weeks, not only in my patients but also closer to home. My wife, Penny, who is my inspiration in most of the "healthy" things I do, was delighted with my transformation from obesity to a slimmed-down specimen of vibrant health. But she continued to eat in her own "healthy" way, which was similar to my style of eating with a few key exceptions: she enjoyed eating "beige" foods like breads and cereals; nor did she take many of the supplements that I would put out for her every day. Despite being a normal weight, she seemed to always have a stubborn 10 to 12 pounds that she was unable to lose. Of greater concern, Penny's LDL ("bad") cholesterol was higher than mine, and she would

Cancel the Vascular Surgery

...

During a routine exam with his cardiologist, a restaurateur and developer named Omar had mentioned hearing a "whooshing" sound in his right ear. The doctor's stethoscope had detected the bruit, a noise made when blood flows through a narrowed carotid artery, in one of two main arteries to the brain. A subsequent ultrasound scan confirmed the doctor's worst fears: Omar's right carotid artery was 90 percent blocked. He referred Omar to me for a carotid endarterectomy to remove the plaque narrowing the artery. After examining Omar, I agreed with the cardiologist's findings and assured Omar that I could do the operation. But I warned him that the problem would recur or manifest in his other carotid artery unless he changed his diet and lifestyle. As a chef, he was intrigued by my food recommendations and agreed to start Diet Evolution.

Meanwhile, a further test revealed that the blockage was in a place that was difficult to reach surgically, which led us to look into other options, such as carotid stenting, which at that time was still experimental. After about two months, a program accepted him for the stent, but he agreed to go only if I accompanied him. Before we left, I did a final check-up. During the interval, Omar had lost about 12 pounds. More importantly, his triglycerides and his LDL ("bad") cholesterol had dropped and his HDL ("good") cholesterol had doubled. Furthermore, the inflammation in his bloodstream had disappeared, and his elevated insulin level had dropped.

I listened to Omar's neck again. Much to my surprise, I couldn't hear the bruit. My patient concurred that he had noticed about a week earlier that the whooshing in his ear had disappeared. We decided to wait a month and recheck. A repeat scan showed the blockage was now only 30 percent, a perfectly safe level, and it remained like that for three years. Unfortunately, Omar began smoking again. At his six-month checkup, I again heard a whooshing sound. A subsequent scan showed that the blockage had returned. Omar quit smoking and redoubled his efforts to follow Diet Evolution. Two months later the blockage was down 20 percent; nonetheless, Omar opted for an operation to allay his fears. When we opened his carotid artery, there was absolutely no evidence of cholesterol plaque; instead, a thin layer of scar tissue was all that remained of the blockage. We placed a tiny patch over it and sewed him up. More than anything, this experience confirmed to him (and me) that following Diet Evolution would be his permanent way of life.

catch colds that I seemed to be able to resist. Then, out of the blue, she decided to test the program, which she knew by heart, having heard me indoctrinate anyone who would listen.

In a mere six weeks, Penny lost 12 pounds, all of it fat! Better yet, her total cholesterol is now 170, her LDL plummeted to 70, and her liver began manufacturing more of the desirable HDL (now at 77) that helps move LDL out of our arteries. Her triglycerides dropped to 45. Even though Penny had always had a desirably low fasting insulin level, it is now less than 1, as low as can be measured. In fact, if I had only one blood test to choose to determine the best shot at longevity, a fasting insulin level is the one I would choose because insulin holds the key to most chronic diseases. As a growth hormone, if elevated, insulin stimulates the growth of cancer cells, thickening of the linings of blood vessels and joints, and acceleration of all aging processes, to say nothing of directing your body to manufacture fat. My advice: ask your doctor for a fasting insulin test, as recommended earlier; then follow my program to get it as low as possible.

SLOW AND STEADY WINS THE RACE

Penny wanted to take off only a dozen pesky pounds, but what if you want to lose a lot more than that? Some of my patients lose all the weight they need to in the first six weeks, but others stay in the Teardown phase for much longer. My philosophy is simple: *do the best you can, with what you've got, wherever you are.* If you keep "beige" and "white" out of sight to the best of your ability, give fruit the boot to the best of your ability, and slow down around "brown," you will eventually slim down. Except for the first two weeks, it's always a good idea to lose weight slowly. And remember, weight loss is only one benefit of Diet Evolution. Fully half of my patients are thin, yet plagued by the same health problems that killer-gene activation brings. All studies of long-term successful weight loss demonstrate that rapid weight loss is rarely, if ever, sustainable.[1] Rapid weight loss is usually the result of short-term control of a radically reduced calorie or carbohydrate intake that can be sustained only for six to eight weeks, before your genetic programming takes control and corrects things. As you'll learn later, it does this by activating hunger hormones that manifest themselves in uncontrollable food urges. No wonder in the past you've read cookbooks like best-selling novels when you've been on a diet. As I've said before, that's why most "fad" diets are written for four to eight weeks of use.

Finally, rapid fat loss is just plain dangerous. Like all other animals, we store heavy metals like lead and mercury and other toxins such as dioxins, PCBs, and organophosphates in our fat cells because they're all fat soluble. That is why women are told to avoid consuming fatty fish during pregnancy—the heavy metals and other toxins in fish could possibly damage the developing fetus. During rapid fat loss, these heavy metals are liberated from the melting fat and enter the bloodstream. Unfortunately, we don't excrete these compounds well in our urine. Our enzymatic detoxification system must inactivate them, but during rapid fat loss the system is overwhelmed, and toxins rise to high levels and remain elevated for a long time. How long? Studies of volunteers at Biosphere II in the Arizona desert, who lost approximately 37 percent of their body mass in six months, showed toxic levels of heavy metals for a whole year before returning to normal.[2] Is this a problem? You bet it is. Dr. Roy Walford, the longevity researcher who headed up the project, died recently from Lou Gehrig's disease, also known as amyotrophic lateral sclerosis (ALS). One cause of ALS is heavy metal poisoning. I'm worried about all those gastric bypass patients who lose 150 pounds in six months. I'd like them to let me know how they're doing twenty to thirty years from now.

So how fast is too fast? Working with the results of the Biospherians, Dr. Walford and his colleagues calculated that a rate of 50 pounds lost per year would allow heavy metals to be safely flushed from the body.[3] Hence my personal planned weight loss of 50 pounds, basically a pound a week, in the first year. Here's my rule: *You just can't beat 1 pound per week!* But again, remember that everyone is different. For some people a healthy and achievable rate may be less than 1 pound per week, for others it may be somewhat more. Also, understand that weight loss is not like clockwork; we're talking averages here. You may lose a half a pound one week and 2 pounds the next, depending on numerous factors.

THE FAB FIVE: SUPPLEMENT YOUR DIET

A multivitamin and mineral supplement from a reputable supplier will ensure at least the bare minimum of 20 to 50 trace elements, vitamins, and minerals that may be missing from even the best diets. Taken twice a day, it forms a good base upon which to restore your "house." The following four supplements also give you the best bang for your buck:

Vitamin E A potent antioxidant, but most products are a synthetic dl-alpha tocopherol, which actually may have no antioxidant activity! Make sure the

Each of us has an individual tipping point, at which the liver converts simple sugars and starches into fat. In fact, by running two successive blood tests to measure triglycerides, which are the first fats the liver forms from sugar, in response to a certain diet, I can predict the point at which a person's intake of "beige" and "white" foods triggers fat manufacture. Moreover, when stimulated by the "Store Fat for Winter" program, your liver manufactures storage-fat transporters, which I believe is the best way to think of LDL. Likewise, think of HDL as a fat recycling truck that circulates through the body and arteries, picking up fat and carrying it back to the liver for reprocessing. My research demonstrates that when you activate the "Store Fat for Winter" program, your liver manufactures primarily LDL and cuts back on its production of HDL.[3] As far as your genetic program is concerned, why send out recycling trucks to pick up fat, if you're giving instructions to your liver to manufacture and store it for the winter?

On the other hand, when you activate the "Winter Is Now" program, your liver assumes it is winter, stops manufacturing fat-transporting LDL, and ramps up production of the HDL recycling trucks to move fat out of storage. Even my normal-weight patients who came to me for treatment of high cholesterol are shocked when they find that making simple changes in their "healthy" eating habits dramatically drops their LDL levels and raises HDL levels.

brand you buy says, "Mixed vitamin E," or if you can't find that, look for d-alpha tocopherols (not dl). Your fish oil supplement may contain some vitamin E as a preservative, but take a separate supplement to be sure to get enough. Contrary to earlier studies, emerging research suggests that for vitamin E to be effective, doses as high as 1,600–2,000 IU may be needed.

Typical daily dose: 400–2,000 IU

Vitamin C Vital to repairing many of the body's systems, vitamin C is also the single most important co-factor for repair of collagen breaks in your blood vessels and skin. If you want smooth arteries, smooth skin, and a reduced risk for sunburn, be sure to take more vitamin C. It also boosts other micronutrients. For example, without it beta-carotene can actually act as a dangerous pro-oxidant instead of a beneficial antioxidant. Animals that can manufacture vitamin C, which humans cannot, produce much more of it during times of stress and infection. When you are stressed, going from one time zone to another, undergoing an operation, experiencing a viral or bacterial infection, or just arguing with your spouse, dramatically increase your intake of vitamin C to bowel tolerance, which usually occurs at 6,000 to 10,000 mg per day.

Typical daily dose: 500–1,000 mg twice a day

Magnesium Most adults become profoundly deficient in magnesium, a mineral essential for muscle contraction and nerve conduction. This mineral is so essential that all heart surgery patients receive 1 to 2 grams intravenously during and every six hours after their operation to normalize their heart rhythm and control blood pressure. Many cardiac disease specialists think a dietary deficiency of magnesium causes or contributes to hypertension. Magnesium is poorly absorbed and competes with the same receptors in the intestines that absorb calcium. When magnesium and calcium are combined in a supplement, you'll actually get less of each. Work your way up to as high a magnesium dose as you can tolerate without having loose bowels (but no more than 1,000 mg a day). After all, the laxative isn't named Milk of Magnesia for nothing!

Typical daily dose: 500–1,000 mg

Folic Acid and Other B Vitamins Mixed B vitamins include thiamin, riboflavin, niacin, vitamin B_6 (pyridoxine hydrochloride), vitamin B_{12} (cyanocobalamin), biotin, and pantothenic acid. Along with folic acid, they were among the first the government required in "enriched white flour." If you're trying to lose weight, I cannot emphasize too much how important high doses of folic acid and B vitamins in general are in reducing the buildup in your bloodstream of the amino acid homocysteine, which has been directly linked to increased coronary artery disease, stroke, and Alzheimer's disease. A word of warning: More than 300–500 mg of vitamin B_6 (pyridoxine) can cause severe neurologic problems.

Typical daily doses: folic acid: 800–5,000 mcg; B vitamins: 50–100 mcg or mg (sold as B-50 or B-100s, which have 50 or 100 mcg and/or mg of each of the B vitamins above)

HOW ARE YOU DOING?

Once you've completed the first two weeks of the Teardown phase, you can make some adjustments in your food choices, which we'll discuss in the next chapter. You've probably lost at least 3 pounds and some of you will have lost more. (The largest weight loss I have seen in two weeks is 12 pounds. The least was a 2-pound gain! Okay, the food diary showed complete lack of compliance.) If you are not losing weight, it's time to start a food diary and write down everything that goes into your mouth. Everything! My Club members who do this inevitably "see" the "white" and "beige" foods that they didn't realize they were eating within the first three or four days of doing this. Younger women often progress more slowly than men, because females store body fat in their hips and butts to allow them to reproduce and nurse their babies. As a result, young women are genetically programmed to hold on to this fat, unlike their perimenopausal friends or menopausal mothers, who fare much as men do on my program. Men also tend to store fat in their gut, where it is more easily metabolized. Gut fat actually regulates normal body weight in most animals by making them feel ill when these cells become engorged with food, resulting in the release of cytokines, which signal the animal not to eat. An ill animal won't eat. But after a few days, gut fat recedes and the hormones are no longer being released, so the animal resumes eating.

READY TO MOVE ON?

Before you leave this relatively restrictive two-week period behind and move on to consuming certain fruits and some whole grains, consider whether you should continue eating the way you have for a few more weeks. To help you make this decision:

- How much weight do you want to lose? *In general I advise clinically obese people, meaning those with a body mass index (BMI) of 30 or more, to continue to avoid fruit and grains until they reach a BMI of 29, putting them in the overweight category. To find out how to calculate your BMI, visit www.drgundry.com.*

- Do you have diabetes or the metabolic syndrome? *I advise my patients with these conditions to stay the course until their sugars and blood insulin levels normalize. Sometimes this happens in as little as six weeks.*

Chapter 7

THE TEARDOWN CONTINUES

At the start of your third week on the Teardown phase of Diet Evolution, you're probably encouraged by your weight loss and bouncing with energy and well-being. Things are different for your genes, too. They're getting the message loud and clear: there are no trees full of sugary fruit around to feast on, and most of your calories come from protein, so it must be the depths of winter. And because you're getting plenty of protein, as well as carbohydrates in all those veggies, your genes are reassured that you're not in starvation mode. You've conveyed these messages by eating as much of the "Friendly Foods" as you want—all without having to bother counting carbs or calories or measuring portions (other than your eyeballing a plate or filling your hand with nuts). Now it's time to make a few modifications—if you wish—and relax into your new eating habits.

You're not going to change the signals you're sending your genes yet; instead, you're going to slowly transition to a new way of eating so that your genes will really get the message about who's in charge. Trust me, convincing them that you and only you hold the key to their survival will activate changes in all your genetic programs that will restore your health and keep you thriving. So for at least the next four weeks, continue to evolve your diet by staying with the "Friendly Foods" listed in Chapter 4 and avoiding "beige" and "white" foods to the best of your ability. In addition, you can make the following changes.

ADD BACK BLACK AND BLUE—AND RED, TOO

Unless you are still obese, you may now begin to add berries (fresh or frozen, but not dried), as well as other black, blue, and red fruits such as currants, cherries, red grapes (just stay away from the seedless types), and plums. These colorful fruits are full of phytonutrients that can help turn on your longevity genes and turn off killer genes. The antioxidants in these fruits are unique in that they are both water and fat soluble, meaning they can travel from the bloodstream to brain cells, with dramatic implications for brain vitality. Give aging rats the equivalent of 1 cup of blueberries a day, and within one month they can navigate through a maze as rapidly as a young rat does.[1] Who says an old dog, ahem, rat, can't learn new tricks? I give my dogs blue- berries! Bears that eat huckleberries live longer than bears without access to them.[2] Want more berries in your life? The first year on my program, I had a bowl of Dr. G.'s Berry Ice Dream (page 262) nightly.

You can also eat up to two servings a day of apples and citrus fruits, as well as tomatoes and avocados, which are botanically fruits, and other "Friendly Fruits" listed on page 69. Continue to steer clear of all "Killer Fruits" (page 67), which tend to hail from the tropics, and fruit juice. (See Tropical Sugar Bombs on page 104.) As long as weight loss continues at a steady pound a week or you're already at your optimal weight, you can enjoy fruit in modera- tion. Overdo it and weight loss will halt or even reverse. If so, simply cut back on portions or wait a few more weeks before trying again.

THE RETURN OF "BROWN" FOODS

You may also add up to a 1/2-cup serving of cooked whole grains or legumes from the "brown" foods list on page 68 per day. But you need not do so. Under- stand that if weight loss is your goal, and it is for most of us, you'll be wise to avoid them altogether until you home in on your goal weight. Grains and legumes are likely to slow or even temporarily stop weight loss. Remember, your ancestors did just fine without them for millions of years. Alternatively, if you can't do without grains and an occasional lentil, cheer up and enjoy them—in moderation, of course. You'll simply lose weight a little more slowly. This slight relaxation on whole grains does not extend to bread, whole grain or otherwise, or any other foods on the "white" and "beige" lists (pages 66–67), which remain off limits. By the way, this is one of the few times I'll ask you to measure food. After doing so a few times, you'll be able to "eyeball" a 1/2-cup portion of grains or legumes.

ADJUST THE RATIO OF PROTEIN TO VEGGIES

As you continue in Phase 1, slowly cut back on the size of your protein portions and simultaneously upsize your portions of "Friendly Vegetables," especially leafy greens. In doing so, you'll decrease the amount of calorie-dense but micronutrient-sparse food and increase the amount of those denser in micronutrients but lower in calories. For example, if you're already eating 2 cups of salad or cooked vegetables a day, up it to 3 cups. If you're already at 3 cups, take it to 4 or 5 cups. If you have always been a big veggie fan, you can eat even more. By the end of your sixth week in Phase 1, your protein servings should be roughly half the size of your palm.

Continue with all other guidelines from the first two weeks, including your supplements, mid-morning and mid-afternoon nut snack, and 8 to 10 glasses of water per day.

See page 175 for a week of meal plans tailored to Weeks 3 through 6.

IT'S NOT JUST ABOUT GLYCEMIC LOAD

If you've studied up on Atkins, South Beach, and Ultrametabolism, you may be convinced that by controlling your glycemic load you'll achieve ultimate health. Once upon a time, I believed this, too, but it's not that simple. Most of the diet gurus who talk about "healthy" pasta, whole-grain breads, and rice conveniently ignore the fact that the populations that consume them without apparent ill effects regularly expend large amounts of energy. If you're working in a rice paddy fourteen hours a day, rice is a superb food source, but not if you're sitting in an office ten hours a day. Just as important, most cultures that use white and beige foods as an energy source have developed ways of dealing with their glycemic impact. For instance, the Japanese serve their rice in a tiny bowl at the end of the meal, so it is released into the intestines slowly, only after the protein and veggies have begun the digestion process. By serving both rice and pasta al dente, or slightly undercooked, the Italians ensure that they can't be digested quickly. Moreover, these grains are served only as a small first (*primi patti*) course. Italians also leave the bread brought to the table alone while waiting for their meal, using it only to sop up the olive oil sauces left in a bowl. They never fall into the habits Americans have developed of consuming mammoth bowls of pasta and endless servings of breadsticks.

The truth is that most of what you have been told about healthy eating and behavior comes from scientists' studies of risk factors. Almost every day on TV you hear a variation of this message: "In a new study, obesity has been associated with a 4.5 times increased risk of breast cancer."[3] This is true, but obesity didn't "cause" the breast cancer. The point is that most of the risk factors are observations of association, not causes. This disconnect I believe is what creates most of our misunderstanding about dietary advice. Here's Dr. Roy Walford's (the longevity researcher who headed up Biosphere II) favorite example: We know that Japanese women are remarkably resistant to developing breast cancer; they also drink a lot of green tea. Many people have made the association between drinking green tea or its active ingredient, a catechin called ECGC, and cancer prevention. But not so fast. Japanese women also all have black hair. We could just as easily make the association that having black hair prevents breast cancer. These silly associations have gotten us very confused as to the real causes of our problems, yet we all blithely believe them as we strive for good health.

After reviewing the food diaries of thousands of my patients in all weight ranges, whose weight and cholesterol or triglyceride levels were *not* dropping, I have found that fruit and what I call "white," "beige," and "brown" foods emerge as the major culprits in blocking forward momentum, regardless of their glycemic index or load.[4] Stop carrying around those silly cards and books with the GIs and GLs of foods; follow my simple rhymes instead.

THE MANY TYPES OF CHOLESTEROL

You may be familiar with "good" cholesterol, known as high-density lipoprotein (HDL), and "bad" cholesterol, known as low-density lipoprotein (LDL). But you probably don't know that there are seven different kinds of LDL, three of which may be dangerous and four of them friendly. Similarly, there are five kinds of HDL, but only one of the five cleans your arteries. Imagine Pac-Man going around gobbling up fat, and you understand how this super HDL works. Despite what you have been told about how important your genes are in determining your risk of high cholesterol, and in how trans fats elevate your LDL, my research has demonstrated that the sugar and starches in foods you eat almost completely determine your cholesterol levels and the ratio of good to bad cholesterol.[5]

The All-American Doughnut Lover

. . .

Mr. All-American in high school, all-state in three sports, and recently honored as one of the all-time best athletes of his state, my friend Jed served in the Marine Corps before becoming a successful businessman. In his mid-sixties, he ran, played tennis daily, and was the epitome of health. But Jed loved doughnuts, Diet Coke, cookies, milk chocolate, and sandwiches made with whole wheat bread. Imagine his surprise when a massive heart attack struck him down at a tennis resort. After inserting a balloon and stent to open the clogged coronary artery, the local cardiologist delivered the bad news: five other blood vessels were severely blocked.

After Jed had recovered, I performed the quintuple bypass and didn't like what I saw. Plaque coated every blood vessel. After the procedure, my associates and I started him on Plavix to keep his blood vessels open, and his cardiologist prescribed a statin to lower his LDL ("bad") cholesterol. I also explained my diet program. Jed made a few changes, such as eating the "heart-healthy" low-fat diet prescribed by the hospital dietician. Six weeks later, his chest pain returned. The stent had closed up and one of the grafts was hanging by a thread. We inserted two more stents, but scans revealed poor blood flow to one-third of his heart muscle.

After being released from the hospital, Jed came back to my office. "Let's start again," he said. He began Diet Evolution that day. I also prescribed supplements to augment his diet—there was no time to lose. Six weeks later, Jed's triglycerides were one-third their previous level. His LDL ("bad") cholesterol followed suit. More important, his HDL ("good") cholesterol, and specifically a subgroup of HDL that actually scours arteries had doubled. Three months later, Jed was skiing down the pristine slopes of Utah's Deer Valley for eight hours a day—pain free.

Two years have passed since Jed's heart attack. A recent scan shows not only normal heart function but also that blood flow has normalized to areas in which it was once poor. Another big difference: He can now run his opponents on the tennis court into the ground.

Pasta Without Guilt

. . .

Can't live without pasta? I've tried every whole-grain, soy, and imitation pasta out there, but I have to thank the Hungry Girl website (www.hungry-girl.com) for turning me onto an important breakthrough in noodle technology: shira-taki. And the amazing thing is that this ancient food has been used in China and Japan for more than 2,000 years. The noodles are derived from the root of a kind of yam, which is finely ground into the most water-soluble fiber known, glucomannan. Oat bran also contains some water-soluble fiber, giving it a lim-ited ability to reduce cholesterol. In contrast, glucomannan can absorb up to 100 times its weight in water! When made into noodles, the fiber is translu-cent and tasteless, but combine it with a small amount of tofu, and it has the texture and appearance of overcooked spaghetti. You can find both the spaghetti and fettuccine styles packed in water in the refrigerated section of Trader Joe's, Whole Foods, and some grocery stores. Make sure you get tofu shirataki, rather than the clear shirataki noodles or you will be disappointed in the texture. Here's the best part: An entire package of shirataki contains only 40 calories, all of it water-soluble fiber, which will reduce your cholesterol, lower your blood sugar, and make you feel full with less food. For recipes made with shirataki, see page 216.

THE VERY WORST CHOLESTEROL

Here's one more example of how our genes use us for their own purposes. About 25 percent of us carry a gene that codes for the manufacture of a partic-ularly small, deadly type of cholesterol known as lipoprotein(a), or Lp(a). Many doctors haven't heard of what I call "bad cholesterol with an attitude," and even if he or she has, your doctor has been taught that neither statin drugs nor dietary changes have any effect on it and therefore they don't bother testing you for it.

If your ancestors hail from northern Europe or the British Isles, there's a good chance you carry the gene. The harsh climate in northern Europe meant that the diet was often deficient in vitamin C, predisposing the population to scurvy. Vitamin C is essential to reweave the broken bits of collagen that routinely occur inside our blood vessels—and our skin. Victims of scurvy often bleed to death from their gums or intestines. Enter Lp(a), which "spackles" holes in damaged blood vessels, allowing our northern European forbears to survive scurvy and reproduce, ensuring the ongoing life of their genes. So far so good. But Lp(a) is such a good spackling compound that it just keeps piling on any damaged area, which means that people with Lp(a) usually develop severe premature coronary artery disease. Why has this gene persisted if it's so lethal? By now you should know. Individuals who carried the gene in vitamin C-sparse areas were able to patch up their leaky blood vessels, enabling them to live long enough to reproduce. Once these individuals had replaced themselves, their genes couldn't care less if they die of a heart attack.

My advice? Get your Lp(a) level measured with a simple blood test. Unlike what your doctor will probably tell you, my research, as reported in the journal *Atherosclerosis,* has shown that Diet Evolution and two simple supplements, CoQ10 and niacin, also known as vitamin B$_3$, will lower Lp(a) to normal levels in the vast majority of people.[6] You can turn off the genes producing Lp(a).

THE TRIGLYCERIDE CONNECTION

The level of triglycerides in each person's blood correlates exactly to his or her intake of "white," "beige," and "brown" foods, as well as of fruits. Specifically, the fruits that consistently get people into trouble are ripe bananas, watermelon, honeydew melons, papayas, mangos, seedless grapes, and ripe pears. Eaten in moderation, those I deem "Friendly Fruits" rarely get people into trouble long term. Just as important, I have yet to find a bread that doesn't raise triglycerides.

If your triglycerides fall, you'll lose weight and your LDL cholesterol will drop. This occurs regardless of any other food intake because you've inactivated your body's storage of sugar as fat. Of course you can lose weight while consuming "white," "beige," and "brown" foods and fruit; I've done so many times. I just ate less of everything. Weight Watchers works this way, as do many other short-term diet programs that may provide initial success. But those pounds inevitably return because all four of these types of food activate our "Store Fat for Winter" genetic program. Cheer up; there is a foolproof way to enjoy these calorie-dense foods without gaining weight: you've got to earn them, as you'll learn in Chapter 11. Meanwhile, during the Teardown phase, *If it's white, keep it out of sight; if it's beige, behave; if it's brown, slow down.*

Tropical Sugar Bombs

. . .

Ripe bananas contain a starch that is rapidly converted to sugar, unlike green bananas or plantains. Cooked plantains, unfortunately, become sugary. Any tropical fruit, including mangos and papayas, has a much higher sugar content than other fruits and, in general, a sweeter taste. You can still enjoy fruits such as bananas and pears, just don't let them get ripe. Eat bananas when the tips are still green and pears when they're crisp. Enjoy green papaya, grated, as the Thais do, mixed into a salad. Seedless grapes, a relatively recent modification, have been bred for a huge pulp-to-skin ratio, making them sweeter than Concord and red wine grapes. The seeds and skins are the beneficial parts of grapes, so stay away from seedless grapes. The same goes for dried fruits, whether apricots, prunes, raisins, or even cranberries. All pack the sugar and calories they had before being dehydrated. Despite the fact that they're marketed as healthy, a handful of dried fruit is the caloric equivalent of three to eight servings of fresh fruit! Just look at the calorie count on a small package of trail mix.

DEMOLISHING THE WHOLE GRAINS MYTH

I touched upon whole grains early in this book, when I introduced the Evolution of the Human Diet chart (page 27). Now that you can choose to include a modest amount of them in your diet, it is time for me to address the issue in greater depth. *Eating calorie-dense foods got us to where we are today*—double entendre intended. Anyone who has followed Atkins, Protein Power, No-Grain, Sugar-Busters, or South Beach Phase 1 diets has been indoctrinated in the disadvantages of grains.

The American Heart Association gives oats the Heart Healthy Seal of Approval, but as president of the board of directors of the Desert Division of the association, I can tell you that the seal is *bought,* not earned. The Scots, known as fearsome warriors, were avid oaters, but they ate whole oats in the form known as steel-cut, Scottish, or Irish oats, which are digested very slowly. Smash an oat grain with a steel roller mill, however, which produces what is known as old-fashioned or rolled oats—or worse yet, quick-cooking oats—and you have a paper-thin membrane that can be instantly digested into—guess what? Sugar. Even worse, grind up that oat into oat flour to make a cheery little "O" out of it, and wham—instant sugar.

Now let's look at the largest crop in the United States. If you grind corn kernels into cornmeal for baking, it too can be rapidly digested into simple sugar, activating the "Store Fat for Winter" program. Today, most corn is dried and then subjected to high pressure and heat, a totally unnatural process that produces corn oil and high fructose corn syrup. Like oats, corn in its original form—whether raw or dried as posole—isn't bad for us, but manipulate it to make it more digestible, and it can do us in.

I know you're skeptical; so was I. You've been repeatedly taught that all grains are good for you, right? This new way of thinking goes against the grain—pun intended—as well as all my training and years of being exposed to advertising. So what's my take on eating whole grains? A truly whole grain may, in fact, be safe for some of us to eat, but it comes at a tremendous price when weight loss is our goal, thanks to its dense concentration of calories. Remember, the density of calories has increased as our diet has evolved. Want to fatten up a cow or a chicken? Feed them grains. Any questions? A 1/2 cup serving of most cooked grains weighs in at between 150 and 250 calories. When starvation was a real threat, grains were a lifesaver.

The same goes for legumes, that huge family comprising pintos, navy, kidney, garbanzo, black, and spotted beans, shelled peas, lentils, and dozens more. All are dense in calories, half of them from sugars, so treat them with extreme caution. Don't confuse legumes with sugar snap peas and green and wax beans, which are immature pods, and are fine in Phase 1. Likewise, black or green soybeans (edamame), pose no sugar threat, although they do have a significant calorie load. Enjoy them but go easy on portion size.

THERE IS FAT—AND THEN THERE'S FAT

At this point, since you're introducing ever-larger quantities of leafy greens and other vegetables into your meals, you'll discover that oil is a key ingredient for turning them into delicious meals. While fat carries flavor and enhances the taste of foods, as with everything in my program, there are also compelling health reasons for consuming fat. But it's important to include the right kind of fats in your diet, which explains why your choice of oil can be a lifesaver. You would have to have been on another planet to be unaware of all

Beware of Artificial Fats

. . .

Not only are the fats in the animals that we eat completely different from what they were fifty years ago, we've also learned to manufacture fats unknown in nature, so-called trans fats, which appear on food labels as hydrogenated or partially hydrogenated oils. Whenever your genes encounter fats they weren't expecting, as in the case of foods devoid of micronutrients, they send you looking for more, assuming that the next bite will have the real McCoy. Only when they can't find real fats will your genes use these fake fats to build components in your cells, especially your cell walls. Imagine your whole body being held together with the equivalent of duct tape. And think about this: If you're eating what your genes perceive as inferior fats, you must be eating lower on the hierarchy, which signals your genes that you're not worth keeping around. So memorize this: *If you eat fake fats, you get heart attacks!*

Trans fats are associated with increases of LDL, but association doesn't necessarily equate with causation. Trans fats are used in processed or fast foods to keep them from spoiling. Both processed and fast foods are primarily grain- or starch-based products. If you eat French fries, chips, crackers, breads, and the like, these starches turn to sugar in your bloodstream and activate the "Store Fat for Winter" program. The underlying grain ingredients are the problem, but the various types of fats keep getting blamed because they happen to be associated with the troublemakers. Avoid these foods and you'll avoid trans fats while getting the added benefit of watching your LDL plummet.

the recent hype surrounding fish oil, omega-3 fatty acids, and olive oil. We have gone from considering fat as the root of all evil to an understanding that our bodies need a certain amount of essential fatty acids, especially omega-3s.

When animals eat green plants, they consume the fats contained in the plants' leaves. This is especially important in fish, which consume significant amounts of algae and seaweed or krill, those almost microscopic sea creatures that are the favorite food of whales. Sardines, anchovies, and other small fish consume sea plants, which are rich in omega-3 oils, concentrating them in their fat stores. When we have a sardine sandwich or anchovies in

Caesar salad, the omega-3 oils get passed along. Likewise, when larger fish, such as wild salmon, eat algae or krill, the omega-3s concentrate in their fat. Further up the food chain, when larger predators like tuna or swordfish gobble up smaller fish that are now loaded with omega-3 oils, they, too, stock up. As the ultimate predator, we eventually eat these large fish and with them the omega-3s.

When you eat fish or take fish or cod liver oil, you're consuming "long-chain" omega-3s. I call all these oils "green" oil. Just as these fish incorporate the green fats they consume into their systems, all animals—and that includes you, my friend—incorporate fats from the things they eat into their own cells and fat stores. The problem arises in that many of the animals we eat today have a radically different diet from what they had in the past.

LOVE THOSE "GREEN" FATS

There are other sources of "green" fats, the most obvious being olives, which when pressed gently as in extra-virgin olive oil yield a "green" oil that contains mostly monounsaturated oleic acid. Avocado is another excellent source of "green" oil. Stop being afraid of guacamole! Just use sugar snap peas or endive leaves instead of tortilla chips to scoop it up. Purslane, a common weed that you probably pull out of your garden and throw on the compost heap, has the highest concentration of the omega-3 fatty acid alpha linolenic acid (ALA)—not to be confused with the omega-6 fat alpha linoleic acid. Don't you just love how we scientists make this confusing?

Linolenic acid is also present in high amounts in walnuts and flaxseeds and their oil, as well as in canola oil, derived from rapeseed, and from hemp seeds—yes, that cousin of marijuana—and their oil. I call these nut- and seed-based fats "brown," as distinct from "green" oils. So impressive are these "brown" and "green" fats, that when patients with coronary artery disease were given a diet containing 30 percent fat, consisting of canola oil and olive oil enriched with extra ALA, and compared to patients who followed the low-fat American Heart Association (AHA) diet, the trial was halted after three years. Why? Because the results for those in the high green and brown fats group were so dramatically better it was considered unethical to have the others continue on the AHA diet.[7] So has the AHA changed its dietary recommendations? No way. But even the ultimate antifat proponent, Dr. Dean Ornish, whose work I greatly respect, now recommends that his patients take daily doses of fish oil.

THE LINK BETWEEN GRAIN-BASED OILS
AND INFLAMMATION

Why are most "green" and "brown" oils so beneficial and "beige" grain-based oils—think corn, soybean, cottonseed, and safflower—so bad for you? As you now know, "green" oils are higher in the essential omega-3 fats, while "beige" oils contain a preponderance of omega-6 fats. Both fats are essential, meaning that although our bodies can't manufacture them, they're necessary for normal cellular function. In general, omega-3 fats reduce inflammation, blood pressure, water and sodium retention, and pain and also relax blood vessels. Omega-6s do just the opposite.

Even skinny folks aren't immune. If you eat an excess of omega-6 fats in the form of "heart-healthy" grain and bread products while limiting your intake of "green" fats, you unwittingly manufacture mostly inflammatory substances with no counterbalancing anti-inflammatory ones. The results appear in the thin women who come to my office complaining of hypertension and joints so sore it hurts to walk or even squeeze their fingers tight. By eating these "beige" fats, they become a bundle of inflammation.

So, how do you restore the balance? It's actually simpler than you think. In the Teardown phase, you have kept the main sources of omega-6 fats—"beige" and "white" foods—out of sight. Evolving our diet back to how people ate about a century ago restores the normal balance of 1:1 between omega-3 and omega-6 fats. Now you understand why I don't limit your consumption of olive oil and keep pushing greens, as well as recommend you eat grass-fed beef, free-range chickens, and wild-caught fish.

CURBING CRAVINGS

There's another benefit to restoring this balance, which is important to your weight-loss journey. When you consume equal quantities of omega-3 and omega-6 fats, your compulsion to eat sweet, sugary food will cease. That's right! It's amazing how many of my patients volunteer this information. In my research, I've found that sugar cravings stem from an imbalance of "green" and "beige" fat consumption and the resultant overproduction of pain and inflammation hormones and underproduction of pain relieving, anti-inflammatory hormones.[8]

Add the antidepressive effects of omega-3 fats, and the urge to self-medicate with sugar disappears. The result: you crave and eat less sugar; the less sugar

Going Green and Brown

. . .

How do you sneak those healthful omega-3s in green and brown fats into your diet? Try these ideas for starters.

• Use primarily extra-virgin olive oil in your salad dressings and drizzle it over cooked vegetables.

• Use extra-virgin olive oil for sautéing. You can also use less expensive superfine olive oil or cold-pressed olive oil—not to be confused with what is marketed simply as olive oil, which is to be avoided at all costs—then finish with extra-virgin olive oil.

• Drizzle flaxseed or hemp oil on salads or cooked vegetables, but never cook with them, which causes them to oxidize.

• Buy your olive oil, fish oil, flaxseed oil, and hemp oil in dark containers. Hemp and flaxseed oils are so unstable that they must be refrigerated, as must omega-3 fats. All rapidly go rancid—even sunlight can trigger this result, as evidenced by a "fishy" odor.

• Keep flaxseed in a tightly sealed container in a cupboard or the refrigerator, and then grind a small handful before serving. If you grind them and store them—even in the fridge—they quickly go rancid. Add them to salads or plain yogurt; make them into granola with coarsely ground walnuts, pistachios, and almonds; add some berries, plain yogurt or unsweetened soymilk, a pinch of stevia, and you've got granola that will cure you, not kill you.

• A teaspoonful of fish oil makes a protein shake smoother and frothier, with no perceptible fishy taste.

• Experiment with drizzling sesame, walnut, almond, or hemp oils over stir-fried vegetables.

• Use a little raw tahini (ground sesame seeds) as a dip for crudités.

you eat, the more the "Winter Is Now" program is activated and fat starts to vanish from your gut or waist. That's why I like to say: *Eat "green" or "brown" oil, and your belly fat will recoil.*

BEYOND BERRIES: SUPPLEMENT YOUR DIET

Blueberries, cranberries, and other berries are powerful sources of antioxidants. But can you do better? Supplementation with cranberry extract, grape skin extract, and grape seed extract seems to me prudent and inexpensive insurance. Want more antioxidant protection? Pycnogenols—combinations of polyphenols from tree bark and alpha lipoic acid—are superb antioxidants, as are mushroom extracts, especially reishi, cordyceps, or maitake. All enhance immune-system function, elevate the level of natural killer cells (another type of white blood cell that protect us), and improve athletic performance. Cordyceps, originally only grown in Tibet and used solely by the Imperial family, were the secret natural performance enhancer used by Chinese Olympic athletes when the nation reentered the games. (For more on these supplements and dosage suggestions, see www.drgundry.com.)

MANAGING HYPERTENSION: SUPPLEMENT YOUR DIET

Hawthorne berries, olive leaf extract, magnesium, and manganese all help dilate (relax) blood vessels. Indeed, many studies suggest that hypertension is a good marker for low stores of magnesium in the body.[9] Most hospital patients admitted for heart disease have low magnesium levels. Magnesium is given to all open-heart surgery patients to prevent irregular heartbeats and potentially fatal arrhythmias as well as to keep their blood pressure down. I start all my hypertensive Club members on regular magnesium tablets, working them up to 500 to 1,000 mg per day. For dosages of the other compounds, refer to www.drgundry.com.

FISH OILS: SUPPLEMENT YOUR DIET

Even if you eat fatty fish every day, you are likely to be dramatically deficient in "green" fats, at the same time you're regularly deluged with omega-6s in grains, fat from grain-fed animals, milk, cheese, and oils made from corn, sunflower, safflower, and cotton. That's why supplementing with fish oils is of huge benefit not only for heart health but also for depression, arthritis, and weight loss.[8,10,11] However, be aware that the content of EPA and DHA, the active ingredients in fish oil, vary widely in different brands. Look for fish oil capsules and bottled oil that is "molecularly distilled," meaning that the heavy metals that concentrate in fish fat (remember, you're what the thing you ate ate) have been removed. And stop wrinkling your nose! Modern fish oil, particularly Carlson's Norwegian Fish Oil, has absolutely no "fishy" taste. I even fooled my wife when I first started sneaking it into our salads; she actually complimented me on how tasty the dressing was. In my opinion, the Trader Joe brand represents excellent value. Can't find a molecularly distilled product? Don't sweat the small stuff: the benefit of fish oil greatly overshadows any threat to your health from the heavy metals. Don't like fishy burps? Odorless or enteric-coated capsules are now readily available. Don't find an excuse not to take this invaluable supplement. A word of caution: don't take fish oils if you are about to undergo major surgery, as they thin the blood. If you are taking blood thinners such as Coumadin, talk with your doctor. Also, no amount of the fish oil will help unless you simultaneously drastically reduce the sources of omega-6s.

Typical daily dose: Start with 1,000 mg and gradually work up to 2,000–6,000

NONFISHY ALTERNATIVES

If you are allergic to fish or for ethical or religious beliefs won't use animal products, flaxseed oil, perilla oil—made from an herb related to mint—or hempseed oil are viable options, but in my opinion they are less effective sources of omega-3 fats. Both hempseed and perilla contain alpha linolenic acid (ALA), which is a "short-chained" omega-3. The two most useful omega-3 fats are EPA and DHA (short for those long, impossible-to-pronounce words I won't trouble you with again), which are "long-chain" omega-3s. Our bodies can join a bunch of short chains together to manufacture long-chains, but it is not clear how efficiently we do this. DHA from algae is becoming increasingly available. Again, do what you can, with what you have, wherever you are.

Hate swallowing capsules or pills? Eat your ALA! Walnuts are a very good source of ALA and represent one of the pillars in Dr. G's World-Famous Nut Mix (page 196). So, for your daily morning and afternoon snack, eat a handful of raw walnuts.

DR. G.'S TAKE-HOME MENU

As a brief refresher course for Phase 1, before we move to a discussion of plateaus and the role of exercise in weight loss, let me remind you that oils from soybeans and corn didn't exist before your parents were born. Further, the more the original food has been altered, the less you should eat it. Finally, the more you extend a food's shelf life, the more it will shorten your life. Now, try to memorize these Gundryisms:

- If it's meat, you get to cheat.
- If it's green, you'll grow lean.
- Give fruit the boot.
- If it's white, keep it out of sight.
- If it's beige, behave.
- If it's brown, slow down.
- When in doubt, leave it out.
- Lose weight fast, say goodbye to muscle mass!
- Weight off fast will never last; weight off slow, you're good to go!
- Eat green or brown oil, and your belly fat will recoil.
- Eat fake fat, get a heart attack.

Chapter 8

SETTLING IN

By now, you're several weeks in to the Teardown phase, and I'll bet you like what you see when you hop on the scale or look in the mirror. Feeling pretty proud of yourself, aren't you? And so you should. You're sticking with the program, adding incremental amounts of leafy green and other "Friendly Vegetables" as you simultaneously reduce the size of your protein portions. And it's working: one by one, those excess pounds are becoming history.

And that's not all. If you once had aching joints, they're starting to feel better—along with the rest of you. If you're subject to headaches or heartburn, you may have noticed that the former are less frequent and/or less intense and the latter has subsided or actually vanished. Those of you on hypertension medicines may be experiencing an occasional dizzy spell because your blood pressure is normalizing and you need to adjust your dosage. Likewise, if you take diabetes pills or give yourself insulin shots, dizziness may indicate your blood sugar level has dropped naturally. In either case, talk to your doctor about lowering your dose and the feeling will go away. Remarkably, some of life's little annoyances seem to have abated. Your coworkers even seem less irritating—amazing how *your* diet seems to be affecting *their* personalities! Even your kids don't seem to be getting on your nerves as much as they used to. Your neighbors and coworkers have started to notice. Life is sweet.

I hate to pop your balloon, but I need to alert you to an almost inevitable hard landing. It has happened to me, to the vast majority of my volunteers, and will probably happen to you. It could happen after four weeks or after twelve weeks, but here it is with no sugar coating: *you're going to hit a plateau.*

It's a Pleasure

...

Sigmund Freud was one of the first to describe our innate desire to find pleasure and avoid pain at all costs. As you've learned, these impulses are hardwired into all animals. When patients embark on my program, they are focused on the tangible markers of success, such as improved cholesterol and blood sugar levels that result from trimming pounds and inches. Certainly, I was. Interestingly, however, the first things that struck both me and my patients—and likely will you, as well—is the elation and absence of aches and pains that follow initial weight loss. You'll simply feel "better" and life's pleasures will come into sharper focus again. Why? Instead of sending you pain signals to get you to stop destroying yourself before it's too late, your genes will be transmitting pleasure signals. Trust me; this is a near universal finding by my patients. It's one of the many "eureka" moments you'll experience when following Diet Evolution. And when you feel this sense of renewal, don't be surprised to find you're suddenly inclined to rearrange your bookcases, give a dinner party, or pick up that tennis racket that's been moldering in the front hall closet. Those activities, in turn, prompt your genes to send still more pleasure signals.

HITTING THE WALL? OR NOT

Your reaction to this sudden slowing or complete cessation of weight loss is probably something along the lines of "Oh great! I thought that Diet Evolution was different from the others I've tried, but I've been eating just as Dr. G says I should, and nothing's happening. This diet is just like all the rest. I might as well give up, because it isn't working." But were you in my office, my first response would be just the opposite: "It *is* working!" So let's assume you're one of my program volunteers and I've ordered new blood tests on you, as I always do six weeks into the diet. When I share them with you, you'll be shaking your head with delight and amazement. Based upon my published research,[1-4] here's what you'd almost certainly see:

- Decrease in total cholesterol of 50 points or so
- LDL ("bad") cholesterol tracking down
- HDL ("good") cholesterol tracking up

- Triglyceride level down
- Insulin level down
- Blood glucose level down
- Blood pressure down 10 to 20 points
- Percentage of body fat down

The diminished triglycerides, blood glucose, and insulin levels alone are evidence that you're inactivating the "Store Fat for Winter" program. You're sending a new set of instructions to your genes and they're responding. Your killer genes have met their match.

THE PAUSE THAT REFRESHES

So, let's assume you've hit a plateau, which I define as two or more weeks of not losing any weight. What's happening? Relax; you've just hit the first of many readjustment spots. I have come to welcome these spots, as much as I welcome a readjustment phase in a yoga pose. Did you know that if runners add a one-minute walk break to each mile run in a distance race, the vast majority actually improve their overall time?[5] That's correct; they cover the distance faster by intermittently going slower. In fact, if you examine any training regimen, the introduction of mandatory rest periods improves performance. Research has demonstrated maximal muscle growth occurs if you rest an exercised muscle for several days between sessions.[6] In Diet Evolution, you're learning a new regimen, a new way of living and eating. I'll explain why you'll hit plateaus, but the important fact to take away from this chapter is that "settling in," a yoga term I'll explain shortly, is critical to your long-term success on and enjoyment of the program.

You'll recall that almost all fad diets talk about losing *x* amount of weight in six to eight weeks. Remember also that almost all of us can follow a regimen based on control for about six weeks, until our genetic programming yanks away control. Were you to resume eating "white" and "beige" foods as well as the small portions of "brown" foods and fruit you have probably reintroduced, you would restart the "Store Fat for Winter" program and suddenly be swimming upstream against the flow of your genetic program. This is why both Atkins and South Beach recommend you return to Phase 1 if you plateau, but it is also why most people get stuck on a high-protein diet for the long run, or more commonly, give up because they miss eating carbs. I have another view of plateaus and a different message to feed your genes.

NO MORE FAT CUSTOMERS!

Here's the real reason your weight loss is slowing down right about now: You've lost fat cells that had been the customers for the food you eat. Your metabolism hasn't slowed down (we'll get to that in the Longevity phase); you simply have fewer fat cells around to "eat" the food you've been ingesting. Sounds odd, doesn't it? But consider this fact: Just to make it through 24 hours, each pound of cells in your body uses about 10 calories. In other words, a 200-pound man requires about 2,000 calories a day just to keep all of his cells well fed; a 130-pound woman requires about 1,300 calories.

So let's review what happened to me eight weeks after I started Diet Evolution at 228 pounds and was down to 213 pounds. When I began I could consume 2,280 calories a day and maintain my weight, but two months later, I could handle only 2,130 calories without regaining. In effect, I'd lost 15 pounds of "customers" for the food I ate, which means that, compared to two months earlier, I had to reduce my intake by 150 calories a day, or my weight loss would slow or stop. Darn! That's two hard-boiled eggs, 2 tablespoons of olive oil, two-thirds of a high-protein/low-carb bar, 1 cup of plain yogurt—you do the math any way you like. Those 15 pounds of cells weren't there to metabolize those calories any more.

Your weight-loss plateau is really just that simple: fewer consumers for the food you consume. So rather than regarding your plateau as a diet failure, consider it as a chapter in your success story. Now here's the good news; when you do hit that inevitable plateau, instead of worrying about getting off it, I want you to spend the next two weeks settling in and holding your weight stable. Continue to eat your nut snacks and add more vegetables, which will provide more micronutrients to help control any cravings for carbs and urge to cheat—and continue to subtract protein. You can also continue to eat your modest amounts of fruit and "brown" foods if you have added them back, so long as you are not gaining weight. If you are beginning to think that I *want* you to plateau, you're right. In fact, plateauing is one real cornerstone of the program. If your first plateau happens in the first four weeks of the Teardown phase, keep doing what you're doing. If it hits in the fifth or sixth week, remain in this phase until weight loss resumes.

There are two reasons for this strategy. The first is perhaps obvious: I want you to learn what it takes to hold yourself at a lower weight. Think about it: Eating as you are right now is just perfect to hold your current weight. Make no other changes in your food choices and your weight will stabilize. This is

When I met Catherine, she was a young mother working in a doctor's office. She casually mentioned that she had been told that she was "pre-diabetic" and had elevated blood pressure, which is like saying you are "a little bit pregnant." A full-figured woman, she sported the classic tummy-overhang associated with individuals who suffer from the metabolic syndrome. Catherine began Diet Evolution, and after losing about 30 pounds, hit her first plateau. During this period, she spoke to me almost daily about her concern that the program had stopped working. I encouraged her to settle in, not to force any more weight off, and to get used to this new weight for a while. She did just that.

About a month later, Catherine was beaming. "The weight loss has started again, just like you said!" She hadn't done anything but maintain her lowered weight. A year and a half and two plateaus later, she is now down 82 pounds, and the pre-diabetes, high blood pressure, and very high cholesterol are history. She wants to trim another 20 pounds and I see no reason why she shouldn't succeed. Her genes want to keep her around.

where most dieters make their first mistake. Since they have stopped losing weight, they assume that the program has stopped working and abandon it for their old ways. Oh well, I'm a failure, this diet is a failure, give me a pound of M&M's! Trust me, you have already learned how to eat and maintain a lower weight. You're a success, not a failure.

LEARNING FROM THE YOGA MASTERS

The second reason for accepting a plateau and continuing to do exactly what you have been doing is less immediately apparent, but perhaps more important. I want you to sit back and observe, to settle in to this new weight. Yoga practitioners get into convoluted poses that go against the natural postures of the human body. In one's yoga practice, there is never a perfect pose; it is always theoretically possible to push the pose "deeper." When you begin practicing yoga, the tension and tightness of your muscles, tendons, and ligaments around your joints actively resist your attempt to achieve a pose. You get to a point where that pose is not only uncomfortable but, if you try to muscle your way past a certain place, the pain or the work required also makes it unbearable. You back off, and so you should, as to not injure yourself. You won't like this experience and may give up on the pose altogether. A good yogini, however, will ask you to stop at the point where a pose starts to get uncomfortable, and relax there instead of fighting to go "deeper." He or she then asks you to focus on your breathing and ignore what your muscles are trying to do. As you settle in and stop trying to resist the tension in your joints and tendons, a surprising thing happens. You actually sink deeper into the pose because you stopped trying. Just like Yoda advised Luke Skywalker, "do or do not, there is no try." This is, of course, the principle of nonresistance that you have been practicing through Diet Evolution. You have done what you can, with what you have, wherever you are.

So, settle in and get comfortable with your new weight—and the new way of eating you can thank for it. Even if you have not yet hit a plateau, take the fifth and sixth weeks of Teardown to observe and learn what has been easy for you in this program and what has been difficult. What foods or circumstances are triggers? What does it take to hold this new weight? Take this time to stop "trying" and absorb what you have done to get here. Start to see where you're going to make the next push. Remember, the tallest, most intimidating mountains in the world are never scaled in a single push: climbers rest and acclimate at staging camps before attempting the next level. Young heart surgeons learn to be accomplished gallbladder surgeons before operating on the heart. Yoga masters did a lot of settling in along the way, and you will, too.

SKIP THE SCALPEL

Need another reason to accept the pound-a-week approach? Try this on for size: If you lose weight slowly, your skin will keep up with it, but lose too fast and you'll almost always need plastic surgery to get rid of excess skin.

MEET YOUR HUNGER HORMONE

Research has shown that this plateau period is characterized by a huge increase in the hunger-stimulating hormone ghrelin, which actually makes you fantasize about how much you want and need food, especially sugars and starches.[7] This is particularly true during the summer months of a diet, when baseline ghrelin levels are at their highest to get you to eat more and store fat for the winter. (Orangutans consume an average of 8,000 calories a day in the summer and merely 3,200 in the off season.[8]) Your genes do this because they're being driven by one of the most primitive forces impacting all living things: seasonal changes in day length and sunlight. In fact, most of my patients have a more difficult time losing weight during the summer because they're trying to swim upstream. Instead, settle in, grab a branch on the side of the river, and hold for a while. When you do this (and you will do this at every plateau you hit), you simply tell your genes that, no, you're not starving to death; it's merely winter and not a lot of food is around, so relax.

While weight loss is often slower in summer owing to this hormonal response to the length of days, you can get around it by limiting your intake of fruit and sleeping more, as I explain below. Also, as you may be likely to exercise more in the summer, you may be able to burn off a few more calories.

SLEEP MORE, WEIGH LESS

There's one more area to explore during our rest period, and that is rest. The media have bombarded us with how we're getting less and less sleep, particularly compared to 100 years ago. The invention of the light bulb allows us to spend much more time awake than in the past, which is great for cramming more into each day, but it's not great for our health. Sunlight controls numerous hormonal and chemical messengers within most plants and animals, a process referred to as circadian rhythm. The amount of sunlight also controls our genetic drive to store fat for the winter when long days start to shorten, signaling autumn's approach. In fact, the clock gene (yes, that's its real name) has been found in all animals, including humans. In summer, when animals are exposed to more sunlight, they sleep less; the reverse happens in winter. All animals follow this pattern, including you and me. Before our ancestors mastered the skill of starting a fire, the setting of the sun meant it was time for bed. Even after acquiring this ability, with fuel in short supply, they didn't light up the night (perhaps we'll relearn that wisdom, one of these days).

It turns out that the two hormones that control hunger and satiety, ghrelin and leptin, respectively, are very sensitive to light and sleep duration. When college students were put in a sleep lab and allowed to sleep for eight hours, the following morning they had high levels of leptin and low levels of ghrelin. The next night they were awakened after only six hours of sleep. This time ghrelin levels were high and leptin levels were low, just as the long days of summer and shorter nights would stimulate us to lay down fat for the winter. But it gets even better: when students were told that they were only going to get six hours of sleep, but then were awakened after eight hours their ghrelin levels were high and their leptin levels were low, despite the extra two hours of sleep.[9] The computer programs for these students was operating on the assumption that only six hours of sleep were in the offing, so their programs sent out the appropriate hunger hormone signals! The lesson: sleep more and your hunger hormone will be less active, making it easier to slim down.

Sleeping more has another benefit. Quite simply, if you're asleep, you can't overeat. Many of my patients eat dinner late and/or are late-night snackers. I know I was. Your ancestors never did this. In fact, as T. S. Wiley and Bent Formby, authors of *Lights Out: Sleep, Sugar, and Survival,* are bold enough to suggest, the onset of chronic Western diseases in a particular group of people can be traced to when electricity use became prevalent in their environment, messing with their circadian rhythms. I wouldn't disagree. As Founding Father Ben Franklin said, "Early to bed, early to rise, makes a man healthy, wealthy, and wise."

I've successfully tricked my genes into thinking it's winter by purposefully sleeping longer in the summer and have lost weight the last two summers. Just coincidence? I doubt it. Hibernating ground squirrels can be convinced to turn in for the duration by manipulating day length, regardless of the temperature in the room. They will happily go into hibernation even in a warm room, as long as the length of time to which they're exposed to light is reduced, simulating winter.[10] Try it yourself. If your genes think winter is here because you're sleeping longer, they'll no longer want to store fat; rather, they'll want to burn it to keep you alive. Just as important, in winter, do what you can to get to bed earlier.

Research shows a direct correlation between length of sleeping time and not only reduced weight but also increased longevity in humans and all other animals tested.[11] Take those ground squirrels, for instance. Their other nonhibernating rodent cousins can live about four years, yet the sleeping squirrels easily reach 22 years of age, a fivefold increase in longevity. Need still more motivation to hit the sack? Similar research has demonstrated that the longer you sleep, the more amorous you and your partner will feel.[12] By now you can guess why. You need to start the process of reproduction as days shorten so you can fatten up just before Junior arrives nine months later, expecting dinner. Birth rates always surge nine months after a major blackout.

WHAT IF YOU'RE GAINING WEIGHT?

When you started the Teardown phase of Diet Evolution, your main purpose was to shut down the "Store Fat for Winter" genetic program by metaphorically removing the "old paint." As you shed old cells, you lay the groundwork for rebuilding your body with modern materials in the form of new cells. But don't be deceived. The "Store Fat" program is always loaded into your genetic computer. That's why you will avoid all "beige" and "white" foods and consume "brown" food in extreme moderation, if at all, in Phase 1. Otherwise, you'll reboot the program and start regaining weight. Eating these foods again is akin to your pulling old boards off a house you're restoring while your buddy hammers them back on.

When I send my blood and that of my patients for laboratory analysis every three months, I play a little mind game. As you know by now, triglycerides are key to determining the level of the "Store Fat for Winter" program stimulation you deliver to your liver. After great strides in weight loss, blood pressure reduction, and control or elimination of their diabetes, many of my patients relax, pat themselves on the back, and let down their guard by enjoying a few more "beige," "brown," and "white" foods. These are the very foods that raise blood sugar levels, provoking an insulin response, which in turn triggers the liver to produce triglycerides, which show up as fat on your hips and gut.

There is a roughly three-week turnaround between the time I draw blood samples to the time we review results in my office. I get to see the triglyceride level just prior to seeing my patient and compare it to previous levels. If the level is down, I guess how much weight the patient has *lost* since I saw him last. Conversely, if triglycerides are up, I try to guess how much weight the patient has *gained,* based on the degree of triglyceride elevation. I'm sorry to tell you this, but I can predict with uncanny accuracy what the patient's weight change will be in those three weeks just by looking at the direction in which the triglyceride level has traveled. The relationship is simple:

- Triglycerides rising = weight gain
- Triglycerides falling = weight loss

As I mentioned before, you can lose weight by severe portion control or severe fat restriction and still eat "white" and "beige" foods. I've done it on the

The Dirty Dozen

...

The following foods are guaranteed to halt your weight loss and tell your computer program that it's time to store fat for winter:

1. "No sugar added" jams, pies, and juices. The ultimate in deception, these are already drowning in sugar, making added sugar unnecessary.

2. Flavored waters or sports drinks. These are almost always full of sugar and/or and artificial sweeteners. If you see the words *sucralose, Splenda,* or *aspartame,* run the other way.

3. High-protein energy bars with the words "sports bar," "energy bar," or "diet bar." These are a red alert. Not to be confused with high-protein, *low net carb* bars, almost all of these are grain-based sugar bombs that hide under the guise of being "healthy" or high in protein.

4. Trail mix or granola. These usually come with the words "all natural," but remember that sugar is all natural, as is cyanide. Both will kill you.

5. Canned vegetable or fruit juices. These provide a serving of fruits or vegetables—along with mucho servings of sugar.

6. Any food that contains sugar using one of its aliases—cane sugar, natural sugar, date sugar, organic sugar, or various syrups. There is no such thing as "safe" sugar.

7. Skim milk and soymilk lattes or Frappuccinos. These have more sugar than the "high-test" versions.

8. Foods such as cereals, breads, and crackers heralding "whole-grain goodness." These might as well be sugar.

9. Foods labeled low-fat, fat-free, no sugar, or no cholesterol. These are all code words for "high sugar" or sweet taste.

10. Diet sodas. All studies show they raise your insulin levels and make you store fat.

11. Fruit-filled breakfast bars or flavored yogurts.

12. Too many nuts. Just how big is that "handful" of nuts you're snacking on? Make sure it's no more than 1/4 cup.

Ornish diet, although my triglycerides shot up to 400! But, whoa, at six weeks, my genetic autopilot took over and my weight ricocheted back! Why? Because I couldn't compete against the activation of my "Store Fat for Winter" program. My genes were convinced winter must be coming and I had to eat if I didn't want to starve to death. My genes won; I lost. (Actually, I gained—weight, that is.)

If your weight starts to rise, I can virtually guarantee you that sugary tastes or foods that quickly turn into sugar are creeping back into your diet. Look carefully at what you're eating and you'll find them. Whenever one of my Club members is stuck on a plateau or even gaining weight, I have the person provide a food diary for two weeks, literally writing down everything that goes into his or her mouth. Several foods with hidden sugars or sweet tastes have consistently reared their ugly heads as culprits. Check out "The Dirty Dozen" (see sidebar, opposite) to see if any apply to you. If you cannot budge from a plateau after two weeks and are not being hampered by any of the Dirty Dozen, you're eating too much protein. Halve your portions and weight loss should resume.

CURB HUNGER AND CRAVINGS: SUPPLEMENT YOUR DIET

You've already learned that omega-3 fatty acids can turn off sugar cravings. Can other supplements help? If you scan diet-supplement commercials, you'd certainly think so, yet very few studies conclusively prove the appetite-suppressive effects of plant compounds. There are a few exceptions, which I've tried on several occasions and measured their effect. Listed in no particular order, all seem to work for me.

St. John's Wort It's the largest selling antidepressant in Germany, and does indeed work on mild to moderate depression by elevating the levels of serotonin in the brain, much like the SSRIs such as Lexapro and Prozac. You'll recall that sugar increases serotonin levels, making us temporarily less depressed; conversely, dieters get grouchy because their serotonin levels fall. When supplementing with St. John's Wort, some of my patients—and I—report diminished sugar urges; others didn't notice such an effect. The only side effects I've seen are two cases of a mild skin rash that immediately resolved upon stopping the supplement.

Dosage: 300 mg three times a day

SAM-e S-adenoslymethionine (no wonder they call it SAM-e), is a compound produced by our cells and not found in abundance in our diet. SAM-e is used in constructing the neurotransmitters dopamine and serotonin, and is catalogued as a mood elevator. It has additional benefits for your joints and liver function, so you get a lot of bang for your buck, but it will cost you up to a buck per dose.

Dosage: 200 mg on an empty stomach daily

Citromax Also called garcinia cambazola, this Brazilian plant also appears to have appetite-suppressing qualities without stimulants. You may find this compound combined with picolinate chromium, which increases insulin's action.

Dosage: 500–1,000 mg twice a day prior to lunch and dinner

For more on these supplements, see www.drgundry.com.

THE EXERCISE CONNECTION

Don't expect exercise to hasten your weight loss in the first few weeks of a diet. Increasing muscle mass can actually interfere with slimming down because it builds muscle, which weighs more than the fat it displaces. (Your clothes will fit better, however, so you may see it in your jeans before you do on the scale.) This is not to say you shouldn't work out if you're accustomed to doing so. And fitness does have other benefits, including psychological ones that can be of particular help if you're stressed from an extended plateau or just need to get your mind off food. Exercise releases endorphins, which have a relaxing effect. Exercise also can help you settle in emotionally at the same time that your body is settling in to your new way of eating. Yoga is an excellent stress reducer, as is tai chi. Although it won't build muscle, meditation is another tried-and-true stress reliever. Remember, your genes are always seeking pleasurable experiences and compounds; mild to moderate exercise not only floods your brain with these compounds, it is actually enjoyable. Strenuous exercise, on the other hand, is rarely pleasurable and sends the opposite message: You are struggling to survive and are not a good example to keep around.

In the next chapter, I'll discuss the fork in the road: the point in any weight-loss program where you confront the fact that you're unwilling to drop below a certain number of calories in a day but want to continue to lose pounds. At this juncture, you have two choices: you have to aerobically exercise to burn more calories, or build muscle through strength and interval training—which is my choice—in order to eat more calories without regaining weight.

MOVING ON?

Perhaps you've lost all the pounds you needed to lose or continue to lose an average of a pound a week. If so, bravo! You have successfully activated the "Winter Is Now" program and are burning fat for energy. Still have some more slimming down to do? You're free to move on to the Restoration phase, but not everyone feels the need to leave the Teardown phase after six weeks. If you're fine with the food choices in this phase and want to hang there longer, be my guest. If you're one of the few people with a weight issue but no health issues (yet), your decision to move on to Phase 2, the Restoration phase, is yours and yours alone to make.

However, if you also have health issues, I advise you see your physician and have your blood lipids and insulin levels rechecked. Specifically:

- Is your insulin resistance gone or nearly gone? If not, stay in Teardown until you have lost at least 20 pounds.

- Has your LDL cholesterol level increased? If so, you're part of a small group of people who have this response to a high-protein, higher animal-fat diet. Time to start rebuilding by moving to the Restoration phase.

One of my patients is a 53-year-old gentleman who has been in the Teardown phase for nine months and lost 72 pounds. You can view his success story on page 75. Yes, he had a lot of weight to lose, and I think he did go a little fast. At this point, he continues to banish about a pound a week, happily eating protein and "Friendly Foods." He's off all his former diabetic and cholesterol medications, and shows no interest in changing what he's doing. That's fine; his risk factors for early death are now miniscule. But to really affect his long-term health, we are starting to evolve him into an eating plan he can live with, literally and figuratively. And I want the same thing for you—a "new house," in the form of a new body. It's time to put away the sledgehammer and start doing some fine carpentry work. Don't worry if you still have pounds to shed, you'll continue to do so, albeit perhaps more slowly, as you shift gears into the Restoration phase.

DR. G.'S TAKE-HOME MENU

- If you "push" to lose weight from a plateau, your genes will push back.
- Enjoy periodic plateaus; all assaults on the summit are done in stages.
- Sleep more, weigh less.
- If your triglycerides are rising, "white" and "beige" foods, or one of The Dirty Dozen, are creeping back into your diet, and you'll start to store fat again.
- If you're fantasizing about food, your hunger hormone ghrelin is sky-high; add more greens and omega-3 fats and it will drop.
- The mind-body connection is real: Get those feel-good hormones activated with exercise, yoga, or tai chi.

Chapter 9

BEGIN THE RESTORATION

Don't think for even a second about blowing off the Restoration phase. Do so and you will not effect permanent changes in your eating habits and lasting success in weight control—or the benefits of improved health and enhanced longevity. Remember, it takes at least three months to make new habits permanent ones. So spending six weeks in Teardown and then returning to your old way of eating will have been an exercise in futility.

In the first phase of Diet Evolution, you followed a way of eating that was as close as possible to the way people ate roughly a century ago, before modern farming methods changed the diet of cattle and other animals from grasses to grains and the introduction of new manufacturing processes that produced grain-based oils and ground grains. During the next six or more weeks, you'll evolve your eating patterns to mimic the diet of our ancestors *before* the onset of agriculture and the domestication of animals roughly 10,000 years ago.

To do this, it's helpful to recall that farming allowed our forbears to grow and store calorie-dense foods in the form of grains, meat, and cheese. While agriculture stimulated population density and led to geographical dispersion, there's no evidence that humankind thrived. And, quite frankly, that's the situation for most of us who eat a diet characterized by "healthy" whole grains and skinned chicken breasts. After studying the effects of a number of foods on my patient volunteers and myself, I believe that the main ingredient missing from our more recent ancestors' diet was green leaves.

Oh no, not rabbit food! Before you throw up your hands and slam the book shut, let me remind you that we share 98 to 99 percent of our genes with gorillas and chimps. A 400-pound male silverback gorilla consumes 16 pounds of leaves a day, but only 3 percent of his body composition is fat. He is pure muscle mass, all generated from protein in leaves. That's right, there's more protein in 100 calories of broccoli than 100 calories of filet mignon. I can just see you shaking your head in disbelief. But consider this: 100 calories of filet would be a thin strip, approximately 1 inch by 3 inches. The whole head of broccoli contains only 100 calories!

And therein lies the key to the Restoration phase. I'm convinced that transitioning from calorie-dense food to calorie-sparse food leads to not only long-term weight control but also to long-term thriving health.

The average bag of pre-washed romaine lettuce contains about 35 calories, evenly divided between protein and carbohydrates. Yes, there is plenty of protein in leaves; just look again at those muscles on gorillas if you don't believe me. Speaking of which—remember I grew up in Omaha—how do you get animals to gain fat rapidly? Not by grazing on the range. Instead, you limit their mobility in a pen, and feed them corn along with soybeans, and they fatten up ASAP, just like us! Limit our activity, feed us a diet based largely on grain products, and we fatten up. Once again, your genes are just responding to the information you're giving them.

Back to lettuce. In contrast, the average apple contains 100 calories, much of it in the form of fructose. You would have to consume three bags of romaine to get the same number of calories as in an apple. Your genes always direct you to foods that supply the most calories for the least energy expenditure. Even if you thought about all the micronutrients that you might be passing up in those bags of lettuce, the shear enormity of eating them compared to the relative ease of munching a single apple drives you to the apple. No wonder our arms were designed to hang from branches to reach fruit! Remember, only the great apes (and you're one of them) have a shoulder joint that allows us to get to the fruit that other primates couldn't reach. But first, let's get one thing straight.

CALORIE SPARSE DOESN'T MEAN "LOW CALORIE"

The term "low calorie" has been used to trick you into eating ground grain products sweetened with artificial sweeteners, which are even more effective in activating the "Store Fat for Winter" program than are high-calorie foods. The confusion between low-calorie and calorie-sparse foods comes down to the nutrient density of foods. In general, most calorie-dense foods have a lot of calories concentrated in a small package. For example, a 1-inch cube of cheese contains up to 250 calories, virtually all of it fat with a tad of protein. (And who stops with one cube, I'd like to know?) What it *doesn't* contain is a lot of plant-based micronutrients (unless the milk came from cows that grazed on grass, but even so, there's not much in 1 cubic inch). Cheese is a calorie-dense, micronutrient-poor food.

On the other hand, consider that half the carbohydrates in romaine are in the form of water-soluble fiber, which means that they're not digested and absorbed as calories, so you can exclude them from the real calorie count. You'd have to consume five to eight bags of romaine to acquire the same number of calories in one tiny piece of cheese. Holy cow! (Chuckle, cow, cheese—get it?) That's a lot of leaves! But those five bags are packed with huge quantities of phytonutrients, many of whose benefits are not yet fully understood. But we do know this: animals that eat green growing things don't get fat, develop heart disease, or become diabetic. Humans who follow a raw and living food diet often have trouble gaining weight. Why? Because the micronutrients and the bulk of plant fiber activate the ultimate satiation hormones produced by cells in your lower digestive tract. That's right, these plant compounds activate the ultimate "I'm not hungry" hormone switch. This switch can stop a rampaging teenager at an all-you-can-eat buffet.

In the Teardown phase, you learned that the more you consumed animal products from animals that ate green things, the healthier you'd get. In each case, the animal was just the middleman between green things and you. To progress with your health in the Restoration phase, it's time to eliminate the middleman. Once again: *If you eat green, you'll become lean.*

How does this happen? First, let's consider volume. Throughout our evolution, humans have always eaten roughly the same amount of food. Our girth is increasing steadily not because we're eating *more,* but because the same volume of food now is more calorically dense. Think about this for a minute, or even try this experiment: Get equal-size bags of potato chips and romaine—they'll weigh about the same. Eat all the lettuce and check how full you feel. Later in the day, eat the chips and see whether you feel equally full. You'll find that both fill you up equally well. But with the chips, you'll have chowed down close to 1,000 calories, much in the form of trans fats, while the lettuce contains a mere 35 calories and is replete with protein and micronutrients.

You're probably thinking that you won't get much satisfaction out of that bag of romaine without salad dressing. Fine. Add several tablespoons of olive oil vinaigrette and you'll still get only 200 calories, but this time the oil will be loaded with micronutrients as well. Any way you add it up, the lettuce leaves are the nutritional bargain and micronutrient powerhouse. You wouldn't use inferior building materials on your home, so why would you put inferior materials in the body that houses you? To reiterate, adding more and more leaves and green foods to your diet will lower your calorie intake compared to an equivalent volume of calorie dense foods. No wonder, *if you eat more greens, you'll fit into your killer jeans.*

GREENS CURB HUNGER

I want to remind you of the second benefit of the micronutrients in green leaves. Your genes have been exposed to plant chemicals over millions of years, so many of our genes are dependent on these phytochemicals for proper functioning and activation. For example, the trace mineral selenium, of which most Americans have dangerously low levels, is intimately involved with controlling a particular gene in your liver, an organ that not only helps detoxify poisons but also regulates the growth or inhibition of cancer cells. Vitamin C, also found in green leaves, is essential to rebuild collagen breaks in your blood vessels and skin. The sun's rays cause wrinkles, which are one kind of collagen break. (Add vitamin C to your face cream and those wrinkles vanish, or at least diminish.) Vitamin C is also essential for recharging other vitamins, including beta-carotene, which becomes a pro-oxidant without vitamin C's help. (Pro-oxidants are the bad guys; antioxidants are the good guys.)

Because these plant compounds have become such an intrinsic part of our cellular functioning, is it any wonder that their presence or absence in our foods would act as a satiety switch? Get enough in your system and you and your genes stop looking for them. Eat foods that are devoid of them and you nosh away, assuming the next bite must contain the nutrients that were always there in the past. The high micronutrient content of leaves and green food provides a double dose of goodness: it gives your body the premium building materials to start reconstruction, and simultaneously reduces your hunger by supplying the required quota of plant chemicals your genetic program is seeking.

A Victim of "White" and "Beige" Foods

. . .

Before coming to see me, 87-year-old Isaiah had undergone two coronary artery bypass operations. In separate procedures, he'd had seven stents placed in various blockages in his heart arteries. Nonetheless, just walking from his bed to the bathroom caused him severe chest pain. No wonder, a scan of his heart suggested that at least half of his heart muscle wasn't getting enough blood flow. Isaiah had been told that additional stents or bypasses were useless. When I reviewed the results of his blood test and dietary habits questionnaire, I immediately found the culprit. Isaiah subsisted on "white" and "beige" foods. Although he was quite thin, he had the characteristic gut of someone who was insulin resistant.

Isaiah started Diet Evolution immediately, and within three months was down 12 pounds with improved blood profiles. His wife, however, was most upset by his weight loss. "Look, he's wasting away," she said. "You're going to kill him!" When I explained the principles of the diet, she agreed to let us proceed. Six months later, Isaiah was able to take walks without pain. At his annual cardiology checkup, a new scan showed normal blood flow to all regions of the heart muscle. Isaiah's wife accompanied him to his next appointment and again accosted me. But this time she flung her arms around my neck and asked me to forgive her for doubting me nine months earlier. "Thanks for giving me back my husband," she said.

I love cardiac surgery. But giving people the tools to heal themselves transcends just about anything I can do with a knife, needle, and thread.

MEET YOUR ANTIHUNGER HORMONES

Greens are packed with fiber, but not the kind you get from high-bran breakfast cereal. Green vegetable fiber increases the speed with which food moves through your intestines. You've been told that fiber is good for you because it prevents constipation. What you probably don't know is that the faster food moves through your lower bowel, the more antihunger hormones in your intestinal cells beam up to your brain, telling you not to eat. I bet you think a gastric bypass (stomach stapling) works by making a person's stomach smaller, right? Wrong. It works by increasing the speed with which food arrives at your lower bowel by putting a short circuit in your intestines that stimulates high levels of antihunger hormones. Even one or two days after gastric bypass, most patients completely lose their previously insatiable desire for food. Gastric bands that merely reduce the stomach size have no effect on these hormones, which explains why weight loss is less effective, and why many gastric band patients rupture their stomachs or continually vomit because of their drive to keep eating.

How do you put this knowledge into action? My research has shown that if you can work your way up to consuming the equivalent of one bag of dark green leaves (lettuce, spinach, or other greens) daily, your life will change dramatically for the better. First, by consuming these important phytochemicals, you send messages that tell your genes how to behave appropriately. Second, eating copious quantities of food, in turn, will activate antihunger hormones—all without taking in large quantities of calories. This relatively low (and effortlessly low) calorie consumption sends a powerful message to your genetic autopilot, saying that you're not a threat to future generations because you're not gobbling up more than your fair share of food. And believe me, your genetic autopilot is always watching.

CUT BACK ON CONCENTRATED CALORIES

The second key to dropping your overall calorie load is to slowly but surely back away from the concentrated sources of calories in our diet, which fall into three categories:

- Meat and other animal protein
- Cheese
- Grains and legumes

Despite being our best friend during the Teardown phase, animal protein (including cheese) should be eaten with caution. I have found substantial evidence in my own practice and other research[1,2] that, in long term, the less meat and animal protein we consume, the easier weight loss becomes and the more we are able to reduce our overall calorie consumption. When it comes to grains and legumes, the caution to eat them in extreme moderation stands. That means no more than 1/2 cup serving of cooked grains or legumes a day.

No, I don't want you to eliminate these foods altogether. And, I promise, you will never have to count calories. Remember, even our primate relatives seem to need at least 6 percent animal-based protein in their diet to thrive—there's that word again. You can certainly survive on a vegetarian and even a vegan diet, and there may be compelling ethical reasons to avoid animal products. However, during my fifteen-year stint as a professor at a medical school that espouses vegetarianism, I encountered very few long-term vegetarians and vegans who were in *thriving* good health. This is largely because the eating habits of most vegetarians in this country make them pasta and grain dependent. They might be more appropriately called "grain-etarians." The vegetarians of southern India, where I perform heart operations as a part of missionary work, develop diabetes and heart disease in their twenties. Repeat, they're vegetarians! Yes, they don't eat meat, but they do eat refined, ground-up legume and grain products and have little access to fresh vegetables. I believe that the key here is really semantics. Rather than thinking in terms of vegetarian or vegan, let me introduce the term *vegephile*: "one who likes to eat vegetables."

Contrast pasta and grain-etarians to raw foodists, whose members include Sting, Woody Harrelson, and supermodel Carol Alt. Raw foodists, most of whom are vegans, don't consume cooked grains or beans and generally have vibrant health for a surprising reason you'll learn about in Phase 3. More on that later, but for now, remember: the more vegetables you eat, the better your health.

LIVE LONGER ON LESS

Want proof that animal sources of protein affect how long you're going to live? Landmark studies of Seventh-Day Adventists, who eat meat less than once or twice a week and consume more than five handfuls of nuts a week, live six to nine years longer than age-matched, "healthy-life style" Californian women and men, respectively.[3,4] But aren't we all living longer? Don't let statistics fool you. Studies show that despite living longer, we are experiencing deterioration in the quality of our health. Surviving is not synonymous with thriving. Check out the nearest nursing home for proof positive.

Is there evidence that eating less meat improves your chances of survival and resistance to the activation of killer genes? Indeed, there is. In the largest study ever done in which individual dietary factors could be isolated, there was a direct correlation between the amount of animal protein consumed and a shortened life span and increased prevalence of chronic diseases that would indicate killer-gene activation.[2] So, dear body restorers, the animal protein that was an integral part of the wrecking crew in the Teardown phase is going to start turning on us if we don't do something different.

IT'S COOL TO LOWER THE HEAT

Why does eating lots of animal protein prove harmful over the long haul? Because the process of breaking down the proteins in any kind of flesh so that your body can use it generates heat. Uh-oh, you're thinking, the doctor has been out in the Palm Springs sun a little too long again. I'll bet you thought that increasing your heat by raising your metabolic rate was a good thing, didn't you? After all, doesn't everyone believe that supercharging your metabolism is good for you? You're supposed to rev up those fat-burning cells, stimulate your thyroid to make more thyroid hormone, goose those fat-burning hormones, and work up a good sweat at the gym! Right? Unfortunately, they're all wrong. To your genetic autopilot, all this means is that you're terribly inefficient at burning your fuel and working much harder than you should. Not worth keeping around, so let's activate those killer genes. Having a high metabolic rate is like having a car that gets 10 miles to the gallon.

Healthy centenarians consistently have temperatures in the range of 95° to 96°F, not the 98.6°F considered "normal," as does my 96-year-old patient Michelle. Let's be clear on this: use the heat of burning protein to jettison those dangerous and unwanted pounds in the Teardown phase, but the sooner you get your metabolism "running" at a lower temperature long term, the better. Even the blood serum levels of thyroid hormone in those thriving senior citizens register on the extreme *hypothyroid* scale, which means they're the animal equivalent of high-efficiency, low-polluting "engines." In other words, these older folks got to where they are by slowly but surely reducing their metabolic rate. This energy efficiency is exactly what your genetic autopilot wants when it sends out signals to reduce muscle mass as you age so you'll eat less. The lower your metabolic rate, the less food you need to eat. Because this means you're not a threat to the rest of the tribe, you get to stick around. Remember those hibernating ground squirrels, with their low metabolic rate, that live five times longer than nonhibernators?[5] Still having trouble with this concept? In an energy crisis, when gas shortages are imminent, which car sits in the garage—the Toyota Prius or the Hummer? Bye-bye Hummer, hello Prius. Do you want your autopilot to keep driving you? If so, pay attention.

FRYING YOUR HEART AND BRAIN

So what does eating less meat have to do with decreasing heat production? Remember how you were so happy to be able to eat all those "free" protein calories in the Teardown phase? Now put your hand on your dog or cat and feel how much warmer it is than you. In fact, the expression "it's a three-dog night" refers to the ability of canines to warm up your bed on a cold night. Do you remember why carnivores sleep all of the time? To drop their abnormally high metabolic rate, caused by the breakdown of animal protein. *If you eat meat, you'll generate heat.*

Here's another reason generating heat is bad: advanced glycation end-products (AGEs), which result from heat's bonding sugar to protein (which we'll discuss more later). Think of them this way: If you want a really crusty steak, you turn up the flame; the higher the heat, the crispier the meat. Unfortunately, the higher your body temperature, the more AGEs form in your brain and heart. You can even see them! Those nasty brown liver spots that seem to spring up overnight on your skin as you get older aren't called *AGE spots* for nothing. Your body is warning you that you're generating too much heat. Instead of heading for the dermatologist's office, make a U-turn to the farmer's market. *Cut down on your meat and decrease your heat.*

This strategy is not about the ethics of eating animals, being the highest carnivore on the food chain, or deforestation of the Amazon to raise cattle. Rather, it is about finding the balance of animal protein to green plants in your diet that is right for you. You may transition to a near total vegephile existence in your Diet Evolution, or you may decide that all you are comfortable with is one meatless day a week. The choice is yours.

CUTTING THE FAT

Many of my patients have spent months on the myriad high-protein/high-fat diets, and most, including me, had wonderful initial results; but at a certain point everything turned around. Here's the other bad news about our wrecking crew from the Teardown phase. The high-fat, high-protein program eventually allowed hunger to get out of control. Remember how you learned that after weight loss the levels of your hunger hormone go sky high, urging you to eat? The bad news for high-fat dieters is that only a diet comprising 15 percent fat (not 5 percent and not 30 percent) and the rest as protein and carbohydrates completely prevents the rise of ghrelin in humans.[6] This finding agrees with those found in the only registry of long-term successful dieters (those who have kept weight off for more than three years). Almost all of these folks, including me, transitioned to a relatively low-fat diet, regardless of their initial weight-loss strategy.

PUTTING IT ALL TOGETHER

So, it's time again to slowly *evolve* your habits: decrease the density of the food and increase the volume of calorie-sparse food you eat so that it speeds through your system, turns off the hunger switch, and lowers your metabolism. This means gradually eating less meat, poultry, fish, and cheese and fewer grains and legumes. Want an easy way to remember this? Most cooked animal protein, grains, and legumes are brown in color. So, *if you want to stick around, cut back on brown.*

At the same time, you're going to find ways to add vegetables, especially green leafy vegetables, to almost every meal. Specifically, you'll aim to eat:

- Bigger, more varied vegetable portions
- Larger salad portions, twice a day
- Significantly smaller portions of all animal protein

For example, if you've become accustomed to eating a portion of meat the size of half your palm, aim to make it one-third the size. The point is that the foods you are eating are basically the same foods you ate in Phase 1, but in Phase 2 you will eat more of some and less of others.

Let's take a moment to get to the real meat of the matter. Our culture has convinced us that we have to have several servings of animal-based protein a day to avoid malnutrition. Nonetheless, despite the fact that most of the people I operate on for heart disease have been doing just that, their blood tests reveal that they're severely protein malnourished. What's up? Sadly, this boy from Omaha has concluded that the obsession with getting enough protein is one of the biggest myths out there. Did the steak you had for dinner last night come from a cow being fed burgers? Of course not! Does Sting look underfed and malnourished? No way. Most animals get all the protein they want or need by eating leaves, and so should humans. What's even more important, if you don't have the plant micronutrient building blocks that your ancestors used to eat (in the form of lots of plants as well as animals that ate leaves), you can eat all the protein you want and still be severely malnourished.

Remodeling Tips

· · ·

As you ease into the Restoration phase, try these ways to remodel meals you were enjoying in the Teardown phase:

- Halve the portion of chicken in your Caesar salad—minus the croutons, of course. A few weeks later, top the Caesar salad with a sliced avocado instead of chicken.

- There's no need to eat an animal (or substitute) protein source at every meal.

- Instead of serving cheese in cubes, scrape off thin slices with a vegetable peeler and serve on a thin apple slice or an endive leaf instead of a cracker, then top with a walnut half.

- Serve a small portion of grilled flank steak or pork tenderloin over a generous portion of steamed asparagus or greens stir-fried with garlic and onions.

- Better yet, serve the meat over a romaine salad or wrap meatballs in lettuce leaves. If you have to have a wrapping, have a high fiber, low-carb tortilla, and then transition to lettuce leaves. Consider this as part of your *evolution* to a healthier, slimmer human being.

- When making an omelet, add snippets of spinach, arugula, or other leafy greens and fresh herbs such as basil, sage, and rosemary. Or serve your poached or fried eggs with sautéed spinach, bok choy, chard, or any leftover cooked veggies.

- If you can't imagine an omelet without cheese, cut back on the quantity of cheese and slowly eliminate it until you have a cheese-free one. Use other "Friendly Vegetables" and seasonings in its stead. I transitioned to sliced avocado instead of cheese and never missed it.

During this phase of Diet Evolution, which you should follow for a minimum of six weeks, you'll gradually advance toward deriving most of your protein from vegetables, as well as nuts and eggs. Remember that 100 calories of broccoli has more protein than 100 calories of filet mignon. As you eat more vegetables, your body will adjust to their high micronutrient and phytochemical content. Continue to eat your nut or seed snacks twice a day and steer clear of

Protein in Nonmeat Sources

. . .

Nuts, seeds, vegetables, and even fruits all contain protein, as this partial list of foods reveals.

FRUIT (RAW)
Grams of protein per serving

	Serving Size	Protein (grams)
Avocado	1 medium	4g
Banana	1	2g
Blackberries	1 cup	2g
Casaba melon	1 cup	2g
Currants	1 cup	2g
Mulberries	1 cup	2 g

MEAT SUBSTITUTES
Grams of protein per serving

Boca Burger	2.5 oz.	13g
Lightlife Ground "Beef"	2 oz.	8g
Tempeh	4 oz.	12–20g
Trader Joe's Meatless Meatballs	3 oz.	10g
Whole Foods Vegan Burger		13g
Yves Veggie Bacon	3 oz.	17g
Yves Veggie Burger		16g

NUTS (including peanuts and soy nuts, which are actually legumes)
Grams of protein per 1/4 cup

Almonds	7g
Cashews	4g
Macadamias	2g
Peanuts	8g
Pine nuts	4g
Soy nuts	10g
Walnuts	5g

SEEDS
Grams of protein per 1/4 cup

Flax	5g
Pumpkin	7g
Sesame	8g
Sunflower	8g

VEGETABLES (All are cooked except romaine lettuce)

Artichoke	1 medium	4g
Asparagus	5 spears	2g
Beans, string	1 cup	2g
Broccoli	1 cup	4g
Brussels sprouts	1 cup	4g
Cabbage	1 cup	2g
Cauliflower	1 cup	2g
Chard, Swiss	1 cup	3g
Collards	1 cup	4g
Corn	1 cob	5g
Kohlrabi	1 cup	3g
Onion	1 cup	1g
Peppers, bell	1 cup	2g
Romaine lettuce	1 cup	2g
Spinach	1 cup	1g
Summer squash	1 cup	2g
Sweet potato	1 cup	3g
Tomato	1 cup	1g

For comparison purposes, a 4-ounce hamburger contains 28 grams of protein; a 4-ounce chicken breast, 30 grams; a 6-ounce can of tuna, 40 grams; and an egg, 6 grams. However, the fat in these protein sources is so much higher than in vegetables that the calorie count is significantly higher.

(Adapted from data in *Beyond the 120-Year Diet: How to Double Your Vital Years,* by Roy Walford, M.D., and *www.ars.usda.gov/nutrientdata*, accessed 8/29/07.)

"white" and "beige" foods. Eat whole grains and legumes in *extreme* moderation (no more than 1/2 cup cooked) or not at all.

SIDE SHOW

Have you ever considered what the term "side dish" means? It's that veggies and grains play second fiddle to the star of the meal, that big hunk of meat or fish of which Americans are so enamored. Now do a mind flip. I want you to start seeing the vegetables as the main dish and animal protein as the side. The more green things you consume with protein, whether meat- or bean- or grain-based, the less of these protein sources you'll naturally eat. You'll simply be too full. So what might a typical day's fare look like? You could enjoy scrambled eggs on a bed of steamed spinach for breakfast, along with a handful of blueberries; a big chef's salad topped with a few sardines or slices of turkey for lunch; and pork tenderloin slices served over asparagus and a side salad for dinner; plus your two nut snacks. For detailed meal plans suitable for the Restoration phase, turn to page 176.

Now that you've become an expert in changing your diet, it's time to burn some extra calories by incorporating fitness into Diet Evolution. A sensible exercise program will boost the already impressive progress you've achieved and keep you on the straight and narrow, even as your hips and rear end get narrower. Read on.

Chapter 10

PICKING UP THE PACE

I'll bet you're surprised that you've reached Phase 2 of Diet Evolution without my exhorting you to join a gym or to run two miles a day. I've deliberately not done so because studies confirm that exercise has little impact on the *initial stages* of weight loss.[1,2] My own and my wife's experiences, as well as that of hundreds of nonexercising volunteers in my practice, further convince me of this reality. On the other hand, all research on successful *long-term* weight loss demonstrates that some kind of exercise program is essential to maintain lowered weight.[3,4] The reason for this takes us back to the cause of your first and any subsequent plateaus in weight loss: you've lost cells that "eat" the calories you consume. Take me, for example; at 155 pounds, I can eat only about 1,550 calories a day. Whoa! That's a bitter pill to swallow. What if I want to swallow more food?

ENERGY IN, ENERGY OUT

You've already learned that one method of eating more is simply to slowly reduce the amount of calorie-dense food and increase the amount of calorie-sparse foods, primarily in the form of leafy greens. But just how low are you willing to go? You'll find out in Phase 3. But for now, if you want to eat more calories, you'll have to earn them as our forefathers did—with physical labor— or they'll be stored as fat. The *type* of physical labor you perform is also critical for telling your computer program just how good a job you're doing, which in turn determines whether your genes think you should be kept around. In this phase of your Diet Evolution, you're going to emulate some of the work your ancestors performed to obtain food. You have several choices, but don't worry; none includes attacking a mastodon with a club or outrunning a gazelle.

MAKE IT A HABIT

It's important to opt for a movement or exercise program that you can develop into a habit. For instance, many months ago, I decided that I couldn't have breakfast until I did pushups and then performed squats while brushing my teeth—I'm now up to 40 reps. No pushups, no breakfast. Simple—and effective. It takes about six weeks to ingrain a habit, and sure enough, a few weeks into this new program as I was brushing my teeth and staring into the mirror, I knew something was terribly wrong. I wasn't doing my squats. Speaking of our work in the OR, one of my teachers, Dr. Mark Orringer, the Chief of Thoracic Surgery at the University of Michigan, is fond of saying, "We do it the same way, everyday; everyday, the same." Take Mark's and my advice: Do it the same way everyday and it will become a habit, just like the rest of your morning routine.

My wife and I do the same thing with dessert. We have to walk about 20 minutes after dinner—you'll find out why the timing is crucial shortly—to "earn" dessert. No walk, no dessert. *To earn it, you must burn it.*

TAKE A WALK

In Part One, you learned that animals move for only two reasons: to find food or keep from being someone else's food. For apes, moving from one dining spot to another means walking on their knuckles. Ouch! Our human ancestors walked upright, which uses less energy, but they clearly walked, not ran, to the next hunting ground. Running consumes too much precious fuel.

Of course they ran when they had to capture a wounded animal or to escape a predator, but if you've ever had a running match with your pooch, you know that our forefathers couldn't have outrun a four-legged animal. (Even my 8-pound Yorkie can do a 5K in 25 minutes—that's an 8-minute mile!) Yet, somehow we have gotten the idea in our heads that we should routinely run for several miles on a treadmill, spin, step, or aerobicize to burn those calories, all in the name of being heart healthy. The long-lived Bushmen in South Africa laugh their heads off at this idea. They know that only *unsuccessful* animals would have to expend such effort. And you know what happens to them!

Beverly, a physician, brought her 38-year-old husband to see me because he kept becoming short of breath while playing with their two children. Chip had gained weight throughout his thirties and was already being treated for high cholesterol, high blood pressure, and acid reflux. Beverly appeared healthy, but suffered from uterine fibroids that had required a hysterectomy.

As expected, despite being on medications, Chip's blood test showed all the evidence of killer-gene activation: high insulin level, signs of diabetes, high triglycerides, and high cholesterol levels, as well as signs of inflammation. Beverly was right to be concerned: the gun was loaded, cocked, and ready to fire! Beverly's labs were much better than Chip's, but her lipoprotein(a), perhaps the most dangerous form of "cholesterol," was very high. Despite being a physician, until then she had never had her Lp(a) level tested.

Husband and wife started Diet Evolution together. I had to wean Chip off his medications for hypertension almost immediately. Six weeks into Diet Evolution, he'd dropped 15 pounds, his triglycerides had plummeted from over 300 mg/dl to 78, his LDL dropped 100 points, and his HDL climbed. His insulin and fasting glucose levels also dropped significantly. When I asked him how he was doing, his reply was typical: "feeling well never tasted so good!" At six weeks, Beverly's weight was down 10 pounds and her Lp(a) had already dropped in half.

Eight months into Diet Evolution, Chip is now off all his medications and 45 pounds lighter. His triglycerides are now 40 and his HDL is higher than his LDL. Beverly is down 27 pounds. One of my colleagues stopped me in the hall recently and asked if I had seen her. "She's a real fox! I wonder what she's doing?" I only nodded in agreement. As for Chip, he recently took his kids hiking in the local mountains at 8,000 feet and ran them into the ground.

SPRINT, DON'T RUN LONG

On the other hand, short, fast bursts of speed would have proved essential when our forefathers needed to catch a wounded animal or sprint to the nearest tree before being gored by a wild boar. Other than that, slow and steady wins the race. My advice: *If you run or walk long, go slow; if you run or walk short, go fast.* There is even more benefit here than meets the eye. I remember when I accompanied my wife, Penny, who qualified for and finished the 100th running of the Boston Marathon in 1996. Looking around the lobby of our hotel in Cambridge, I thought that there must be a convention of cancer survivors at nearby Harvard Medical School. Pale, with wasted muscles, and generally sickly looking, all these elite marathoners looked just like the cancer patients I was treating at my local hospital. As I've since learned, running marathons is a catabolic sport, meaning you lose muscle mass dramatically. When my colleagues at Loma Linda studied the marathoners, they found that their immune systems were shot. In contrast, look at sprinters in the 100- or 200-yard dash. They're muscle bound—exactly the opposite of long-distance runners. *If you run sprints fast, you'll gain muscle mass.*

HEAVY LIFTING

Our ancestors were clearly into another form of exercise: both men and women collected food, whether prey, leaves, berries, or tubers, and carried it back to a central camp. Today, we call it strength training, but in the old, old days without such activity you and your tribe didn't eat. This is another way of saying: *If you lift weights, you will lose weight.* And for goodness sake, forget about lifting light weights with lots of reps. What kind of nonsense is that? Your ancestors lifted heavy things! Next time you go grocery shopping, skip the cart and instead grab a couple of hand-held baskets. Not only will you have to set them down and pick them up every time you need to get something off a shelf, but by the end of the shopping trip you'll be carrying your gathered goodies back to camp (actually the car), just like your forefathers. Your genes will be delighted.

PUTTING SAYINGS INTO ACTION

What does all this mean for you in practical terms? Our ancestors lifted, pulled, wrestled, and sprinted after things; they also walked long distances. The more you duplicate these actions, the more your genes will identify your behavior as that of a successful animal. The more muscle mass you carry into old age, the more you must still be hefting food back to camp or fighting saber-toothed tigers, and therefore valuable to the tribe and your collective gene pool. In contrast, the more you behave like a struggling animal, con-

. . .

If you walk 10 or 20 or more minutes after a meal, you'll drop pounds faster than if you walk the same distance before the meal. Why is this? Walk after a meal, and your monitoring system senses that you're heading for the next campsite or hunting ground. It doesn't know whether you're going to walk one mile or twenty. Because of this uncertainty, it doesn't make sense to store the food you just ate as fat, as you may need all those calories on your trek. On the other hand, walking before the meal sends the opposite message: you've arrived in camp with the goodies and you aren't going anywhere, sending your genes the message that it's time to store fat for later use. How does a computer beat a master chess player? It knows patterns of behavior that precede and follow each move and plays accordingly. We acknowledge this ability in a computer, but have a hard time imagining such sophistication at our molecular level. Your genes have been programmed to behave this way for millions of years.

stantly running, jazzercising, and eating micronutrient-poor food, the more you tell your computer program you're not worth keeping around.

What's more, remember that original imperative for all animals? Find food for the lowest calorie expenditure. What in the world would your genes think about your getting on the treadmill or stepping on and off a step? They'd tell you to quit! Why do you think all that expensive exercise equipment sits in the corner and that gym membership goes unused? Your genes are not stupid; they made you stop. As long as you act like an animal whose genes are worthy of preserving, they will protect you, but regularly subject yourself to constant pain, as severe exercise does, and it's a different story. Your genes view pain avoidance as one of the three cardinal rules of survival and will try to get you to stop a painful activity. Consistently ignore these warnings and the second tier of killer genes activates to get rid of you and your genes.

On the other hand, weight lifting, followed by a hunk of high-quality calories rich in micronutrients, is another story. My daily pushups and squats, followed by a small breakfast, imitate our early ancestors' behavior. (See "No-Pain

No-Pain Workouts

...

Here's a ten-minute workout that will let you live like a primitive man or woman in the privacy of your home. In addition to pushups and squats (essentially deep-knee bends), try these basic moves:

- *Raising the jug:* Using a pair of gallon jugs (drink the water first), refill them with water to whatever weight you can handle. Sitting in an armless chair with your legs planted on the floor, experiment with lifting the jugs from the floor along the sides of the chair or in front of you, first to eye level, then to above your head. Add enough water so that you can do it only three or four times.

- *Leg lifts:* Lie on the floor with a rolled towel placed under the small of your back. Lift one leg and then the other about 12 inches off the floor and hold it as long as you can. If that's too intense, do one leg at a time. Repeat until it's impossible to hold them up. It won't take many.

- *Situps:* While you're down on the floor, finish up with as many bent-leg situps as you can manage. Increase the number as you get stronger.

- *Biceps curl:* With your arms at your side, raise one jug and rotate your forearm until it is vertical and your palm faces your shoulder. Lower to original position and repeat with opposite arm. Continue to alternate between sides. Keep doing this until you can't lift any more. If this is more than five or six times, add more water to make the jugs heavier.

In about ten minutes you'll let your genes know that you're working to keep them fed. So, after this workout, you get to eat something!

For more information on life-lengthening exercise, I recommend *The Power of Ten,* by Adam Zickerman, and *The Slow Burn Fitness Revolution,* by Fredrick Hahn.

Workouts," above.) I've evolved my morning routine to imitate and stimulate my ancient genetic computer program without a whole lot of extra effort. Can't do forty pushups? Neither could I a year ago. But the fifteen pushups were just as hard then as the forty are now, and so is the effort expended and imprinted on my genetic program, which worked just as well then as it does now.

THE INSULIN CONNECTION

There is one more benefit to building muscle mass and decreasing body fat: reducing insulin levels. Insulin "sells" food to muscle cells, but as your muscle mass usually declines with age, insulin has fewer and fewer customers to which to make a sale. As a result, insulin has to work harder, and your pancreas pours out more insulin to "push" food into cells blocked with fat. Whew, what an effort. It's no wonder you're tired all of the time. But as you begin to rebuild muscle mass, you start building the customer base for insulin. When you exercise those muscles, their cells start screaming for more food. My wife's experience is a perfect example of how this works. Penny's insulin level has always run a nicely low 3 or 4 (normal is less than 10). But when she finally adopted my diet and reduced her running program, replacing it with walking as well as weekly sessions of strength training and yoga, she dropped 12 pounds in eight weeks, all of it body fat, and her insulin level plummeted to less than 1. Lowering insulin levels keeps the "Store Fat for Winter" program from activating, but more important, it also eliminates the possibility that insulin will stimulate the growth of cells that don't need stimulating.

MUSCLING UP: SUPPLEMENT YOUR DIET

You'd have to have your head in the sand not to know that steroids increase muscle mass. Elite athletes in baseball, cycling, and track and field have all been implicated in their use. But beyond the media hype and public outcry, an important fact is overlooked: you can also stimulate muscle growth and power with supplements, including the two that follow.

Coenzyme Q-10 CoQ10, for short, is critical for muscle strength and stamina. One of the most ominous trends in modern medicine is the widespread prescribing of statins to lower cholesterol. Statins work by limiting the action of an enzyme called HMG-CoA reductase in the liver, which is also responsible for the manufacture of CoQ10. In our rush to lower cholesterol levels, rather than looking at the cause for elevation, we have created a population with dangerously depleted levels of CoQ10 in muscle cells. The consequence of this enzyme blockade may take up to a year to develop, but if you have muscle pains or weakness or have developed congestive heart failure, consider supplementing with a daily dose of CoQ10. This agent is also one of the only nutrients other than niacin that can lower a particularly dangerous type of cholesterol called lipoprotein Lp(a).[5]

Typical daily dose: For anyone on a statin drug, take at least 50 mg; to reduce Lp(a) levels, at least 150-250 mg is useful

Acetyl-L-carnitine or L-carnitine These act at the level of the muscle fiber to transport energy into and around the individual muscle cells, particularly the energy-producing mitochondria. Studies of my patients with congestive heart failure and/or cardiomyopathy have often shown dramatic improvement in heart muscle and other muscle function with daily supplementation.[6] Two of my patients improved to such an extent that both were removed from the heart transplant list. All of my cardiac surgery patients now receive this supplement postoperatively, which is also available in prescription form under the name Carnitor.

Typical daily dose: Either 125-250 mg of acetyl-L-carnitine or 250-500 mg of L-carnitine, twice a day

MOVING ON

After you've spent at least six weeks in Phase 2—assuming you are getting close to your goal weight and your cholesterol and other markers indicate that you're continuing to make progress—it's time to think about moving to Phase 3. In the next two chapters, we'll look at how to extend not only your life span but, more important, your "health span."

DR. G'S TAKE-HOME MENU

Memorize the following Gundryisms, and you'll be well on your way to acing the Restoration phase of Diet Evolution:

- If you cut down on meat, you'll reduce your heat.
- A lot of greens plus a bit of meat make a meal that can't be beat.
- If you eat dark green, you'll become lean.
- The more you eat greens, the better you'll fit into your killer jeans!
- To earn it, you must burn it.
- If you lift weights, you'll lose weight.
- If you run long, go slow; if you run short, go fast.
- Sprint fast, and you'll build muscle mass.

Chapter 11

THRIVING FOR A GOOD, LONG TIME

Over the past twelve or more weeks, you've transitioned from Phase 1 of Diet Evolution, which emulates the diet people ate roughly a century ago, to Phase 2, based on the hunter-gatherer lifestyle of our earlier ancestors. As a result, you've normalized your weight—or are on a steady course to do so—enhanced your health, and are well on the way to making permanent changes in your lifestyle. Phase 3 reaches back even earlier for inspiration.

I regard the Longevity phase as the natural culmination of my program. But I am well aware that it is not for everyone. Up to now, you have been eating both cooked and raw foods, the latter primarily in the form of salads. In Phase 3, you eat primarily raw food, as our earliest ancestors did. For them, the opportunity to consume meat and other animal sources of protein was not a daily event; they relied instead on plant protein, consuming most of it raw. Eating food raw preserves more of the micronutrients, although there are exceptions—notably the lycopene in tomatoes, which is more bioavailable in cooked tomatoes. Another benefit of raw food is that it is considerably more bulky than its cooked counterpart. A bag of spinach reduces to just a couple of serving spoonfuls when sautéed. All that raw bulk makes you feel satisfyingly full so you naturally eat fewer calories.

I find that the idea of eating raw food is more appealing when you think of it as partaking of choices at a giant salad bar. Raw spinach may not appeal, but a

More Than Skin Deep

. . .

Your skin acts as a barometer of what's happening inside your body. As you proceed through Diet Evolution, you'll almost definitely notice a difference in your skin. One of my coronary bypass patients also suffered from diabetes and hypertension, until I convinced him to practice Diet Evolution, which he has now done for the past two years. He recently returned from his 55th high school reunion and reported that his female classmates kept touching his face and asking him what kind of skin-care products he used. They told him that his skin looked like it did when he was a teenager, whereas the other men looked like they were dying. It was at this point that he realized that his body had evolved and the killer genes had been shut down and his old, worn-out cells tossed overboard. His genes were now protecting and preserving him as their most valuable asset. A kid at 73!

spinach salad garnished with pine nuts, baby tomatoes, and a crumble of blue cheese is quite another matter. Speaking of nuts, in this phase you continue your twice-daily nut and seed snacks, but cut the portions from 1/4 cup to 1/8 cup (2 tablespoons) to reduce the calories. To learn how to eyeball this, grab your usual handful and then put half the nuts back in the bag and the other half in your tummy.

Why move to yet another phase? You may feel you've already won the prize you were after. Certainly, the changes you have already made will impact the quality of your life and likely your longevity. But I suggest that now you have an opportunity to enter the bonus round, with its exciting nuances on the cutting edge of the science of longevity. So let's do this: I'll tell you how I have further evolved my diet and you can decide if some or all of it appeals to you. As I have said all along, do what you can, with what you have, wherever you are. My objective is not to lay down the law, but rather to provide you with information and options.

Before we get to the specific ways in which you take your diet to the next level so your genes know that you deserve to live a long, long life blessed with robust health, you now know me well enough to not be surprised that first I have to explain the scientific underpinnings. Some of this is a bit complicated, but I'll make it as simple as possible, so please stay with me. It's vitally important for you to understand the "why" so that you can concentrate on the "how" that follows. Specifically, you need to understand the role that hormesis plays in activating your longevity program. *Hormesis* is the generally favorable response of an organism to low exposures to toxins and other stressors that in large doses would produce the opposite effect. At the risk of sounding like a broken record, hormesis can best be explained by Nietzsche's observation: "That which does not kill us, makes us stronger." Or as I like to say, what is "bad" for you is actually "good" for you.

SURVIVAL IN THE DESERT

All plants and animals possess subtle sensors that alert genes of approaching hard times so they can take measures to protect themselves. Key to your longevity program is the self-protective response that kicks in when scarcity threatens. To set the stage, let me tell you about the desert plants on the San Jacinto Mountains outside my window. All living organisms need water to survive and to reproduce, but Palm Springs and the rest of the desert Southwest don't get much rain. Our plants have developed a strategy to cope with the sudden and unreliable appearance of minute amounts of water every spring. Dormant most of the year, desert plants bide their time, essentially in a state of suspended animation in which their energy demands decrease but they become incredibly resistant to insects and other infestations.

With the first appearance of rain, the plants initiate an accelerated process to shoot out leaves to manufacture energy and produce flowers that can be pollinated and produce seeds, and then go dormant again. Thanks to their genes, which take advantage of good times and rapidly make lots of genetic copies, in a matter of weeks these plants accomplish an entire year's worth of growth and reproduction. Desert plants that have evolved to accomplish this remarkable feat of reproduction survive in this inhospitable environment. The plants' genes act just like all other genes but in an exaggerated fashion when stimulated by water. The corollary to this is that during a dry season, desert plants sit it out. Last spring, when rains failed to arrive in Palm Springs, the plants protected themselves by staying dormant, waiting for better times.

Inherited High Cholesterol Is a Myth

. . .

Stu and Sally were in their mid-40s, a little chubby despite being "healthy" eaters and regular exercisers—Sally practices yoga and jazzersize five days a week and Stu hikes. Both had high cholesterol, but because they ate healthy and were active, their regular doctor assured them that their high cholesterol was inherited. Therefore, he told them, they would both need to take cholesterol-lowering medication. They came to see me hoping that a change in diet might help solve the problem instead.

After their first two weeks of Phase 1, Sally was gloating! She had banished 9 pounds and Stu was right behind her with 8 pounds. A great start. Three months into the program, and sure enough, all their cholesterol problems were history. Both are 20 pounds lighter; but more important, they've learned that they didn't inherit bad cholesterol genes. They've just told their cholesterol genes to "be good!"

A LITTLE STRESS GOES A LONG WAY

Hormesis actually *improves* resistance to infections, tumors, and death. For example, in difficult-to-believe yet completely reproducible experiments, mice subjected to low levels of radiation throughout their lives lived an average of 30 percent longer than unexposed siblings.[1,2] Yes, instead of killing or weakening mice, a low dose of radiation actually made them live longer. Other experiments involving environmental stressors such as heat, cold, lack of nutrients, ultraviolet light, and toxins all come to the same startling conclusion: at the right dose, these potentially lethal factors can actually promote survival.[1] Just like desert plants, animals that survive environmental challenges are naturally selected to reproduce when times get better.

I take advantage of hormesis when I perform heart surgery by briefly shutting off the flow of blood to the heart. This period of stress alerts the patient's heart cell genes that trouble is on the way, activating a complex series of events that cause heart muscle cells to hunker down and protect themselves until a better time—in the form of new blood flow—arrives. Even better, other cells in the area that are not pulling their weight are eaten by white blood cells or instructed to die, a process called apoptosis. The result is that strong cells remain and weak cells are eliminated, enhancing the chance of survival during and after heart surgery.

All the research on hormesis in humans suggests that there is a point in every exposure to a stressor or toxin in which the response is not only to survive but also to thrive. That's because your genes realize both that it is not the right time to reproduce and put offspring in harm's way and that unless they can keep you well, you may die, taking your genes with you. A case in point: Moderate exercise is a known hormetic stressor that results in increased life span,[3] but too much exercise can result in a shortened life span. Whoa, there it is again. *Moderate exercise is good for you because it's bad for you.* Because of this stress response, some exercise improves your immune system and too much destroys it. The bottom line is, whenever your genes sense too much stress, they conclude that you're not a successful animal and therefore don't need to be protected and preserved, activating killer genes. But sense a little stress, and your genes keep killer genes at bay.

EAT LESS, LIVE LONGER

Another example of hormesis occurs with calorie optimization. Eating just enough calories while getting plenty of micronutrients significantly extends life span in all creatures tested, including worms, fruit flies, mice, rats, dogs, and most recently, rhesus monkeys.[4] How much of an increase? In certain species, life span increases up to a staggering 600 percent. Labrador retrievers lived four years longer when their calories were restricted by 25 percent after the age of 3, compared to littermates fed the "normal" amount of calories.[5] But if calorie optimization is so good for you, why are stress hormones such as cortisol and epinephrine (adrenaline) elevated in animals on low-calorie diets? Only now are researchers beginning to understand that calorie optimization is a hormetic stressor that activates an animal's ancient genetic programming to increase its chances of survival under adverse conditions.[1,4,6] Limiting calories makes the animal's genes circle the wagons to protect and preserve the species. Genes send signals to the animal to hunker down until more food appears, just as desert plants bide their time, waiting for a wet spring.

No matter what type of food they eat, animals that consume fewer calories live longer. But what about exercising away those extra calories that we want to eat, as I suggested in the last chapter? Mice on low-calorie diets that don't exercise look exactly like mice that consume more calories but burn them up with exercise. However, the calorie-restricted rodents still live longer than those that work out regularly on their mouse wheels.[4] The evidence is over-whelming: *If you consume fewer calories, you'll live longer.* People who live to be 100 or more typically lose weight progressively after age 30, as their muscle mass diminishes, until they get to roughly what they weighed at about age 13, right before puberty reared its hormone-wracking head. Typically, such long-lived individuals eat about one-third less than they did at age 30.

WHY VEGETABLES ARE GOOD—AND BAD—FOR YOU

Have you ever been curious, particularly if you're a parent, why it is so diffi-cult to get your children to "eat their vegetables," but when they approach adulthood suddenly they "acquire" a taste for zucchini et al.? Our genes pro-tect our rapidly growing childhood cells from these plant compounds in the easiest way possible: they turn on or off taste bud receptors. During the years of rapid growth, your taste buds say "yuck"; when the danger is past, and you might need some help getting rid of rapidly growing cells that could manifest as cancer, your taste buds are activated to say "yum" instead!

There is also considerable evidence that the nausea common in early preg-nancy is a protective mechanism to prevent women from consuming too many plant compounds during the early months of fetal development, when critical cells and organs are being formed.[7] As a congenital heart surgeon, I know that all organs are fully formed two and one-half months into pregnancy. Isn't it amazing that most of the nausea of pregnancy suddenly resolves at about ten weeks' gestation? Coincidence? No, our autopilot is merely protecting the next generation of genes from plant toxins.

Hormesis explains why people who eat a primarily vegetable diet tend to be shorter, reach menarche later, and live longer.[8] The stress of battling plant poisons causes genes to send out a signal to lie low until things improve. All vegetables have hormetic properties, but bitter vegetables seem to have par-ticularly strong ones. Populations that favor bitter foods, as the Japanese and Italians do, are noted for both their longevity and their short stature. Want to activate hormesis with food? It's simple: *More bitter, more better.*

Now that you're an expert on hormesis, it's time to put all this knowledge into action. Let me tell you how to get started on Phase 3 of Diet Evolution.

Is Detoxing Toxic?

...

If you swap your Western diet for one composed exclusively of raw vegetables and juices for even a short period of time, you're likely to experience what natural health practitioners call detox symptoms: headaches, skin rashes, fevers, chills, and lethargy, to name a few. But the real reason you experience these symptoms is not the result of purging your body of built-up toxins. Rather, the symptoms result from a sudden massive exposure *to* plant toxins, which your liver enzymes can't process fast enough. The detox diet is actually a toxin diet. How do I know this? I treat numerous individuals who are about to undergo or have had stomach bypass/stapling surgery. They may rapidly lose 100 to 150 pounds, with their shrinking fat cells releasing enormous amounts of heavy metal "toxins" into their bloodstream, but I have yet to see a single one develop "detox" symptoms. Likewise, the scientists of Biosphere II lost one-third of their body mass in a year and had huge elevations in heavy metals measured in their bloodstream without any "detox" symptoms.

ACTIVATE YOUR LONGEVITY PROGRAM

By now you're already well on your way to being a vegephile. You've restored your body by using top-quality "building materials" in the form of at least a bag of pre-washed leafy greens a day, along with other vegetables. You're getting most of your calories—and most of your protein—from vegetables. You've developed a habit of exercise that you can maintain. You're looking and feeling better than you have in years and are heading toward your optimal weight. Now, to evolve eating habits that will prolong your life and ensure that you thrive no matter what your age, here's how to induce hormesis.

GIVE YOUR STOVE A VACATION

Take the easy steps first. You're already centering your diet on vegetables, which are inherently calorie sparse, so you're effortlessly consuming fewer

Anne is the striking wife of one of my physician patients, who came to see me in "secret" after witnessing her husband's transformation into a green eating machine. A "health nut," she seemed to be doing everything right. She consumes at least as many supplements as I do and is a fitness fanatic, but now that she is approaching middle age, she had a stubborn, slowly creeping weight gain of about 15 pounds that, despite her best efforts, would not go away. She long ago severely restricted grain products; she eats "organic." What's left? Starve?

Anne had done everything right, and at 5' 5" and 136 pounds, her weight was well within the normal range, but her body fat percentage of 32 tended toward the high end of normal. Anne's problem stemmed from two factors. As she approached middle age, she had already lost at least 20 percent of the muscle mass that she had in her twenties. Moreover, she had fallen into the protein-is-an-animal trap. Like most Americans, she was convinced that you obtained protein from animals or animal by-products—end of discussion. When she learned about our gorilla genetic soul mates, she was off and running for the produce section. "I love salads! I thought they had to be topped with a skinless chicken breast to be healthy," she said. Anne was able to start right in on Phase 3 of Diet Evolution, which she is now happily following.

In three months Anne lost only 12 pounds, but reduced her body fat by 4 percent. By adding strength training to her workouts—lift weights, lose weight—and backing off animal protein and loading up on micronutrients in greens, she's almost at her new goal. Not only that, but her HDL cholesterol is 95, up from 50, while her LDL is 60, down from 95. Her insulin level is an exemplary 2! Not only is she delighted, her husband now gives me a mischievous smile every time he sees me in the hall.

calories than you used to without reducing the amount of food you're eating. As I just discussed, calorie optimization induces hormesis, as does exposure to plant toxins. But plant toxins are reduced or inactivated by cooking. As part of your Diet Evolution, I want you to consume even more phytochemicals than you have up to now, so I'm going to ask you to gradually stop cooking some of your vegetables and leafy greens and eat them raw instead—and to continue this practice for the rest of your life. Add raw foods to your diet slowly to avoid experiencing any toxic responses such as headaches, rashes, diarrhea, and joint aches that many people mistake for the side effects of "detoxification."

To begin your exploration of raw foods, follow these suggestions:

- *Ease into eating raw vegetables by cooking some favorite dishes for a shorter time. Get used to the taste and texture of veggies that are par-boiled, meaning they are cooked so briefly they're still crisp. Or, instead, stir-fry them in extra-virgin olive oil for two minutes, then sprinkle on some toasted sesame oil. Wow! Talk about a taste and texture explosion.*

- *Make crudités, the French word for "raw vegetables," your munchies of choice. String beans, broccoli and cauliflower florets, zucchini spears, sugar snap peas, celery, and countless other veggies are great on their own and for dipping into guacamole, hummus, or other dips. Or try my Seed-Sar Salad dressing (page 198) on any combination of raw vegetables.*

- *Broaden your definition of a salad by adding raw vegetables such as thinly sliced raw beets, cauliflower, bell peppers, and snow peas to your leafy greens. A chopped salad of celery, red onions, peppers, radishes, cucumbers, and tomatoes is a treat when you just can't face another let-tuce leaf. In the recipe section, I provide several recipes for salads, but feel free to use your creativity. The possibilities for tasty salads are vir-tually endless.*

- *Use shredded cabbage, grated zucchini, or bean sprouts in lieu of rice or noodles in Thai or Chinese dishes, like my Angelic Jungle Princess with Chicken (page 244).*

Don't stop with vegetables. Sear tuna or salmon and serve it rare. When dining in a Japanese restaurant, order sashimi, not sushi. If you have a fishmonger who sells sushi-grade fish, you can make your own sashimi. Gravlax, the raw cured salmon Scandinavian delicacy, is another gourmet choice. Enjoy beef or lamb carpaccio on a bed of arugula leaves. How about steak tartare? Numerous cuisines developed such delicacies in the days before refrigeration. Do use a quality butcher and fishmonger, and make sure that they understand you are planning to eat their products raw. Don't, however, risk consuming raw poultry or pork.

BITTER IS BETTER

Now is a good time to broaden your vegetable repertoire. If you always grab a bag of spinach and ignore the bitter greens, how about going for kale, chard, beet, collard, or mustard greens, or even dandelion leaves, instead? All take well to stir-frying with garlic and a splash of olive oil and lemon juice. Young dandelion leaves, arugula, and watercress add bite to a mixed salad or can stand on their own with an assertive vinaigrette. Radicchio adds color and pungency to any salad. Red and green cabbage are both great raw, as is Chinese cabbage, especially topped with chopped cashews and roasted sesame oil and cider vinegar.

EAT EXTREME GREENS

There are those who love the taste and texture of seaweed, but I'm not one of them. However, I'm so convinced of the power of algae and seaweed that I now supplement my diet with capsules and tablets of Klamath Lake blue green algae, spirulina, chlorella, and red and brown marine algae, and other green supplements.

But for those who do like their flavor, go for it! Whether you call them seaweed or sea vegetables, these greens (actually, they come in an array of greens as well as red, black, and brown) are wonderfully easy to prepare. Nori, the shiny black sheets used to wrap sushi, can also be used to wrap tuna or chicken salad, sliced avocado, or other "sandwich" fillings. Or cut nori into thin strips to top a salad or soup. Numerous other kinds of seaweed can be rehydrated and tossed into soup or over a salad or soaked before draining and dressed with oil and vinegar—try sesame oil and unseasoned rice vinegar. You can also mix in other raw vegetables such as grated carrot or daikon. There are dozens of varieties, but those you're most likely to come across include hijiki, dulse, wakame, limu, and laver.

THE RAW TRUTH

I promised you earlier that I would tell you my current eating pattern. But first, let me backtrack a little. When I first experimented with this phase of the diet, I ate all my food uncooked for two months before sending my blood samples to my testing lab. The results were the best I'd ever had. My total cholesterol was 170, my LDL 70, and my HDL 77. All my inflammatory markers were at almost immeasurable levels, plus my insulin level was 2. No, my gray hair did not turn brown over night (after all, male silver-back gorillas eat raw food and have gray hair), but my skin tone and elasticity changed dramatically. I was so convinced by this experience that for the following year, I ate about 95 percent of my food raw. In developing the recipes for this book, I backed down to 90 percent without any noticeable changes in my skin, energy levels, or blood tests. Will I return to nearly 100 percent? I doubt it. But I will tell you that after even a couple of days on the lecture circuit away from my own kitchen and favorite restaurants, my cravings for raw food, especially salads, become intense.

My most successful patient volunteers, meaning those who permanently change their body and blood chemistry, do so by evolving to a diet that is at least 60 percent raw food. Should you eat mainly raw? That's your decision. Again, do what you can, with what you have, wherever you are.

In a typical day, I might eat the following meals:

Breakfast A green smoothie made from organic fresh or frozen vegetables that I alternate every few days, using rice, hemp, or whey protein powder, combined with a cup of frozen berries, a whole small apple, flaxseeds, and some cinnamon. Alternatively, I'll have a handful of raw nuts and an apple or handful of berries. In a pinch, I'll have a high-protein, low-carb bar.

Lunch A huge dinner plate of romaine or other lettuces, cherry tomatoes, whatever other veggies or mushrooms I can find, olive oil, vinegar or lemon juice, and crushed red pepper flakes, plus a handful of mixed raw nuts, primarily walnuts.

Dinner A typical meal at our favorite Italian restaurant in Palm Desert recently included sliced raw artichokes with olive oil, fresh lemon juice, and shaved Parmigiano-Regianno cheese, followed by a plate of thinly sliced beef carpaccio topped with arugula, olive oil, lemon juice, and capers. The night before, I pureed half a bunch of raw asparagus with some plain yogurt and spices while briefly stir-frying the other half. I mixed these with a bag (okay, it was two bags) of shirataki tofu noodles, topped with a few shavings of Parmigiano cheese and freshly grated pepper for a delicious "spa-ghetti" meal. For dessert, I had a piece of greater than 70 percent cocoa extra-dark chocolate with a sprinkling of raw cocoa nibs, which makes it like the best crunchy chocolate bar you've ever had, but this one's good for you. And for the finishing touch—one our early ancestors never got to enjoy—an espresso!

For a full week of meal plans and recipes suitable for Phase 3, see page 178.

After recently being invited to a panel discussion on biomarkers in aging, I was impressed by research data regarding the effect of fasting every other day on longevity-gene activation (a technique you'll learn about in the next chapter). I will be testing my biomarkers after this book goes to press, but to find out how I'm doing, follow my progress at www.drgundry.com.

Warning: I am a professional Diet Evolutionist on a closed course! Don't attempt this technique early in Diet Evolution, or you may find yourself flat on the floor with a very low glucose level.

In the next chapter, the last before the Meal Plans and Recipes section, I'll introduce you to more ways—in addition to your diet—that you can employ to induce that funny-sounding but life-lengthening state called hormesis in Phase 3, aka, the rest of your long, healthful life.

Chapter 12

TRICKING YOUR GENES: BEYOND DIET

As you've progressed thorough Diet Evolution, I've advised you to eat more vegetables, particularly raw vegetables, to increase your intake of micronutrients. You've also learned about and practiced many of the principles of calorie optimization, especially in this phase of the program. But we now know that its benefits come not from actually consuming fewer calories, but from the *stress* of consuming fewer calories. This means that many alternative styles of eating—taking an occasional break from the standard three squares (and two snacks) a day, as just mentioned—can also induce hormesis and thereby activate your longevity genes.[1-4]

As with the suggestions in the last chapter, you may be inclined to go all the way with me, you may decide none of these ideas is for you, or you may wind up somewhere in the middle. (I do suspect that you will be interested in drinking red wine and coffee and eating dark chocolate!) Wherever you land, I hope you will give some of these options a whirl:

GET ON THE FAST TRACK

· *Fast every other day and eat two days' worth of food on alternate days. Do this once a week or make it an every-other-day pattern for the rest of your life, whichever works for you. I prefer to fast on Thursday. Think of it this way: fast Thursday, and then enjoy a virtually guilt-free weekend. Fast in a way that works for you. I drink only water, coffee, and tea, but if that's too hard for you, diluted lemon juice and/or vegetable juices (but not fruit juices) work as well. Fasting also refers to simply eating a whole lot less than you normally would.*

- *Skip a meal, starting with one day a week, and then work up to every other day. The more often you do it, the more of a hormesis response you activate.*

- *Occasionally, skip breakfast and lunch and eat all of your calories at dinner or eat only one meal a day.*

In fact, a recent Utah study showed that following this last technique just once a month dramatically reduced the development of coronary artery disease in already very healthy Mormons.[4] Every time you try one of these options, you'll activate the hormesis response. But please be realistic. Just as you shouldn't be heading down Black Diamond ski runs on your first day on the slopes, these techniques are useful after you have evolved your eating patterns over time. Also, don't abuse these tools in an effort to lose weight.

Conventional wisdom holds that skipping meals is bad for you, but conventional wisdom has made you ill, fat, and headed for an early grave. The wisdom of Diet Evolution is based on the science of how our genes, your autopilot, interpret the information you give them in your meals and behaviors. Do you really believe that a million years ago, your great uncle Zeke wiped the sleep out of his eyes every morning and then picked up a bone to gnaw on or a leaf to chew because he knew he had to have a hearty breakfast? Come on! He'd have to gather or hunt for his "breakfast," and if he didn't find it until evening or the next day, that's when he ate. His genes were designed to make sure he could make it until then without food; otherwise, Zeke—and his genes—wouldn't survive. Zeke was probably very familiar with the concept of fasting. If you want to get back to nature, eating three meals a day at a set time is about the most *unnatural* thing you can do.

A TOAST TO YOUR HEALTH—IN MODERATION

Plants are not the only source of toxins. How about the poison ethanol, found in all forms of alcohol? All studies that look at the effect of consuming alcohol over time show a classic hormetic curve, meaning some is good and lots is just the opposite.[5] Individuals who drank small quantities of alcohol—meaning one or two servings a day for women and two to three for men—were found to live longer and have lower rates of heart disease than people who abstained or consumed alcohol rarely as well as those who overindulged.[6] A recent 14-year study of nonsmoking doctors who were of normal weight found that those who consumed two glasses of red wine a day virtually eliminated the chance of a heart attack over the study period, while those who were teetotalers actually had numerous heart attacks![6] But have more than one to two (for women) or two to three (for men) servings a day and the toxic properties of alcohol raise blood pressure and promote cirrhosis of the liver and eventually cardiomyopathy.[7] By servings, I mean a glass of wine or a 1-ounce jigger of spirits.

In low doses, alcohol stimulates your blood vessels' endothelial cells to manufacture TPA, the same compound given intravenously to dissolve clots after a heart attack or stroke. Moreover, alcohol stimulates these same cells to produce nitric oxide, which keeps blood vessels from constricting. That's why you look flushed when you imbibe. (Believe it or not, that's what Viagra does as well.) But here's an important proviso: If you don't drink alcohol, don't start. If you do drink, be aware of the steep risk-to-benefit ratio. Alcoholism is a real disease; alcohol hits the same pleasure center in your brain that sugar, cocaine, tobacco, and sex stimulate. Alcohol is powerful stuff, so use it wisely.

Red grapes produce a phytochemical called resveratrol, which protects them from ultraviolet radiation, fungus infections, and other stressors. Resveratrol activates anti-aging genes that stimulate the production of proteins that circulate in your body, throwing overboard cells that aren't pulling their weight and rejuvenating cells damaged by advanced glycation end-products (AGEs). But don't rush out to the health food store to buy resveratrol capsules, which are generally worthless because resveratrol is inactivated by exposure to oxygen. The fermentation process that produces wine yields carbon dioxide, which protects the resveratrol from the effects of oxygen. Interestingly, the higher the altitude at which grapes are grown, the more resveratrol they produce to protect against sun damage. In Sardinia, where grapes grow at 4,000 feet above sea level, there are more centenarians per 100,000 people than almost any other culture. As I like to say, *if you drink red wine, you'll be fine!*

COFFEE, TEA — OR CHOCOLATE?

Whether or not you enjoy a glass of wine with dinner, I suspect you do like chocolate. If so, here's some really good news. The phytochemicals, particularly EGCG in cocoa, act much the same way that other plant toxins do and their consumption similarly follows a hormetic curve. Does that mean you can gobble up milk chocolate guilt free? No way. The active ingredients in cocoa mimic the bitter polyphenols found in green tea and coffee, but all are completely inactivated by milk. That's why only very dark chocolate, as indicated by 70 percent or more cocoa content, has a beneficial effect. Raw cocoa beans and non-Dutch processed cocoa also contain these polyphenols. You can add them to coffee to make mocha or even vegetable smoothies, stir-fries, and sauces. The ancient elixir of the Incas and Aztecs, mole sauce, was based on cocoa. Or try my recipes for dark cocoa smoothies and chocolate ice cream using raw cocoa nibs mixed with dark chocolate. Yum!

Likewise, enjoy your coffee and tea minus milk, or use unsweetened plain, chocolate- or vanilla-flavored soymilk if you want to preserve the phytonutrients. (West-Soy and Trader Joe's make good unsweetened soymilk.) Now you know why the British, who generally add milk to their tea, don't enjoy the same health and longevity benefits as the Japanese, who drink it straight.

Both black and green teas, as well as coffee, have plenty of phytochemicals, but interestingly their benefits are released only in the presence of caffeine. So unless you have an issue with heart palpitations, stay away from decaf. Recent studies have nearly completely absolved coffee or tea as causing long-term health issues; in fact, it is becoming clearer that both of these beverages have health-promoting properties so long as you don't drink more than five cups a day.[8]

BLOW HOT AND COLD

Here's another fun way to activate longevity genes—with heat. Exposure to higher than normal body temperatures for even short periods of time activates the production of compounds known as heat-shock proteins. These remarkable proteins tell any cell that is not carrying its weight to self-destruct, leaving only fresh healthy cells. These remarkable proteins actually make your cells impervious to damage. So my advice is to go to a sauna or a steam room; take a bikram, or hot, yoga class; or come visit me in Palm Springs in the summer. Being exposed to cold also activates those longevity genes. Scientists studying the responses of hibernating animals to the stressor of subnormal temperature—as well as depleted energy stores and diminished oxygen supply—have isolated a compound that activates cell protectors in hibernators—another example of hormesis at work. Growing up in Omaha and Milwaukee, and living in Ann Arbor, Michigan, for many years, I knew that the weather "toughened" you, but now I know why. In fact, the Swedes, despite a diet high in salt and meat, have the best health and longest life span of any industrialized country.[9] It's cold up there! If you live in a climate with cold winters, get out and enjoy winter sports or just a walk in the cold. If you live in a warm climate, maybe you can learn to love the occasional cold shower. Both extremes of temperature signal your genes to take up a defensive posture. That's why hibernating animals tend to live longer—cold activates longevity.

PUMP YOUR MUSCLES

As you already know, exercise is hardly the best way to lose weight, but as you've also learned, muscle mass increases the calories you can burn. Following the exercise plan described in Chapter 10 will add just the right amount of stress to your system. Again, exercise prolongs your life because it is stressful, not because it is beneficial in itself. I know it sounds crazy, but now you know the real reason exercise is "good" for you is that it's "bad" for you.

TAKING A BREAK

If you use some or all of the tools I've described in this chapter to induce hormesis, you'll arrive at a very interesting place. Remember *Groundhog Day*, the movie about a conceited local weatherman, played by Bill Murray, who relives February 2 seemingly forever in Punxsutawney, Pennsylvania? He awakens every morning as if nothing he did the previous day had happened. He is immortal, but it's always the same day. One night, drinking in a bar, he turns to two drunken men and asks them what they'd do if they were trapped in the same place, every day were the same, and nothing they did made any difference? Their conclusion? We could do anything we want! What this means for you is that, once your genes have received the message that you must be preserved and protected, thanks to the hormesis you've induced you can, within reason, do anything you want.

Well, you can and you can't. First, if you have followed the plan, you have arrived at the point in Phase 3 where you've normalized your lipid and glucose levels. If you eat that piece of chocolate cake when your insulin level is still high, it will go straight to your hips and undo everything. My frustrated patients (yes, I do get a couple now and then) are the ones with very elevated insulin levels who initially do well and are so pleased that within weeks they resume eating bread or pasta or having a bag of cherries daily, and their weight loss stops or reverses. They had not lowered their insulin levels enough to "get away with it"; when these "Store Fat for Winter" foods arrived in their body, insulin was more than happy to turn them into fat.

Second, such departures from the straight and narrow are not license to run hog wild. Eat anything you want day after day, and pretty soon you'll be back to having your genes rule—and undo all the good work you've accomplished in the last three months. But can you take an occasional break from your new lifestyle? You bet!

- Want to have a slice of pizza, a baked potato, or a sandwich? Okay, but skip a meal the next day to make up for it.
- Want to pig out at a barbecue? Don't eat the day before or make sure you fast the day after.
- Heading to a big dinner party? Eat little or nothing most of that day.
- Did you overdo the alcohol on vacation? Back off for the next week.

As long as the overall trend of your diet and lifestyle challenges your body with low levels of stress, your health, your resistance to infection, and your vitality will improve. And as long as you are consuming the optimal number of calories, you won't regain weight. Sounds like a plan!

WHY VITAMINS ARE GOOD—AND BAD—FOR YOU

How about the antioxidants, vitamins, minerals, and other anti-aging supplements I've recommended? Many studies have suggested that in high doses, antioxidants may turn into "pro-oxidants,"[10] meaning that they actually promote the oxidation that can damage organs and stimulate the aging process. This research fuels the ongoing debate about whether vitamins and other supplements are good for you. At a certain dose, almost all vitamins, minerals, and other supplements show toxic side effects. Recent evidence suggests that antioxidants like vitamins E, D, A, DHEA, and melatonin are alike in their effects in that they show exactly the same hormetic response curve as any other stressor.[10-13] At low doses they're good for you and at high doses they are bad for you. You could even say that low-doses of supplements induce longevity, and high-doses of supplements decrease it.

RUN COOL, RUN LONG

Studies of animals and humans that are examples of successful aging show that decreased body temperature is a common feature associated with longevity.[14] As you've learned throughout this book, heat is a true instigator of shortened life span, not only from its direct effect on the production of AGEs but also as a signal of inefficient energy use, as in miles per gallon. True energy efficiency—think of yourself as a hybrid car instead of a gas guzzling V-12—translates into more years lived for the amount of food consumed. Your genetic calorie counter is always computing your fuel consumption and comparing it to what's allocated by your computer program for you to grow up, reproduce, and get out of the way so as not to compete with your own offspring or the offspring of others.

As I've alluded to earlier, the fewer concentrated sources of protein you consume as you move through the phases of Diet Evolution, the "cooler" your metabolism runs. The fewer calories you give your system to work with, the more efficient your cells become. If you follow the phases I've outlined for you, I feel confident that you can achieve what most of us want: to die "young," at a very old age. Or as Mr. Spock would say, "Live long and prosper." My last "success story" of Pansy and Mark should serve as inspiration to stick with the program for the rest of your long, healthy, smooth-skinned life. I'll look forward to hearing from you with your own success story before long.

DR. G'S. TAKE-HOME MENU

To live a long, vital life blessed with good health, memorize these Gundryisms:

- If you eat less, you'll live longer.
- Eat food "live" to arrive at a hundred and five.
- Vegetables are good for you because they're "bad" for you.
- Exercise is good for you because it's "bad" for you.
- Drink some red wine, and you'll be fine.
- Keep your genes guessing as to the timing of your next meal.
- The cooler your engine runs, the longer you'll go without a major breakdown.

Many people ask me whether there's a point of no return, when it's too late for an individual to restore his or her body. Whenever this question arises, I tell them about Pansy, who was 80 when I first met her, and her husband Mark, a relative kid at 70. When I moved to Palm Springs, I took up tennis. I wasn't very good, but I would show up at the Palm Springs Tennis Club every Sunday for a game and was invariably teamed with Pansy, a thin, slow-moving lady. But she had game. That is until one day when she couldn't chase down any tennis balls, and she told me how playing made her short of breath. A subsequent echocardiogram and then a cardiac catheterization confirmed the worst: most of Pansy's coronary arteries were closed or blocked and her mitral valve was severely leaking. Shortly thereafter, I operated on her, repaired her valve, and performed five bypasses. Sadly, I couldn't perform a ventricular reduction (removing the damaged pieces of the heart and rebuilding it), for which I am famous because too much of her heart was dead.

Pansy recovered well and was soon toddling around the tennis court, but about a year later she stopped coming. Shortly thereafter, Mark brought Pansy back to see me. She could barely breathe, her legs were swollen, and her mental function, to be charitable, would be described as Alzheimer's-like. Mark had gotten into an argument with her cardiologist, who had guaranteed them that this was the best that could be expected and that the end was near. Mark assured me that they were ready to try anything.

Despite being an avid and excellent tennis player, Mark carried the telltale "beer" gut signifying that killer genes were activated in him as well. Sure enough, when we tested the two of them, Mark was insulin-resistant, hypertensive, had high LDL ("bad") and low HDL ("good") cholesterol levels, and high triglycerides. Pansy's results weren't much better than her husband's. Their genes were literally killing them. I assured the couple that if they both went on Diet Evolution, we could turn things around, and they agreed. An examination of their eating habits revealed the typical Western diet, helped along by huge amounts of "healthy" rice, since Pansy is Asian. Out it went! In went my dietary program and a host of supplements.

In three months, Mark was down 25 pounds, his insulin levels had fallen, and his blood sugar—once 111—was now 95. His HDL had skyrocketed and his blood pressure had returned to normal. Two years later, and another 10 pounds slimmer, Mark demolishes his tennis opponents, which has made me unpopular with them.

As for Pansy, we ditched most of the standard heart-failure medications she had been prescribed and started her on two different ones, plus high-dose acetyl-L-carnitine and CoQ10. She has became a "green" eating machine. She initially lost 10 pounds of water weight, with the result that her blood pressure looked like that of a healthy teenager, and started to become more active. A year later, Pansy plays tennis daily and has joined a group of stronger players. She has gained 10 pounds of muscle weight. Her hair is thick and vibrant, and her eyes twinkle. Her mental functioning is sharp as a tack. Who says you can't teach old genes new tricks? We made Mark and Pansy's genes work for them, not against them! And so can you.

Part Three

MEAL PLANS

AND RECIPES

MEAL PLANS

In the following meal plans, dishes indicated in boldface type can be found in the Recipe Index on page 280. A few other points:

- *Follow the appropriate recipe version for your phase.*

- *Dress salads with extra-virgin olive oil and vinegar or lemon juice, or use one of the dressings in the Recipe section. Dress vegetables as well with extra-virgin olive oil.*

- *If possible, use omega-3 eggs in omelets and other dishes.*

- *After the first two weeks, if you wish you can add back tomatoes and avocados, as well as one or two daily servings of fresh or frozen berries or fruits that are not "Killer Fruits" (page 67).*

- *For dessert, have berries or a serving of "Friendly Fruits" (after the first two weeks; page 69), a piece of more than 70 percent cocoa chocolate, or see recipes for desserts in the Recipe section.*

While you are becoming accustomed to eating the Diet Evolution way, you may find it handy to copy and enlarge the meal plans for the phase you are following and post them in a handy location such as a bulletin board or the inside of a kitchen cabinet.

SAMPLE MEAL PLAN, PHASE 1: WEEKS 1 and 2

	MONDAY	TUESDAY	WEDNESDAY	THURSDAY	FRIDAY	SATURDAY	SUNDAY
BREAKFAST	Basic Omelet	High protein/ low-carb shake	2 eggs scrambled with low-fat cottage cheese	Yogurt with Nutty Granola	Dr. G's Koffi-Mocha Freeze	Mushroom Omelet	High-protein/ low-carb bar
SNACK	¼ cup raw nuts/seeds	¼ cup raw nuts/seeds	¼ cup raw nuts/seeds	¼ cup raw nuts/seeds	¼ cup raw nuts/seeds	¼ cup raw nuts/seeds	¼ cup raw nuts/seeds
LUNCH	Seed-Sar Salad topped with tuna	Caesar salad (no croutons) topped with chicken, tuna, or anchovies	Hamburger patty wrapped in lettuce leaf Raw celery, carrots, peppers	Caprese Salad with sardines	"Dream" of Broccoli Soup Spinach salad with cubed turkey breast	Tricolore Cobb Salad	Arugula, red pepper, and onion salad topped with ham
SNACK	¼ cup raw nuts/seeds	¼ cup raw nuts/seeds	¼ cup raw nuts/seeds	¼ cup raw nuts/seeds	¼ cup raw nuts/seeds	¼ cup raw nuts/seeds	¼ cup raw nuts/seeds
DINNER	Chicken breast Roasted Cauli-flower with Sage Arugula salad with mushrooms	Fettuccine Dr. Gfredo Green beans. Tossed green salad	Simply Grilled Alaskan Salmon Broccoli Boston lettuce and watercress salad	Turkey breast Brussels Sprouts You'll Eat Spinach salad with mushrooms	Alaskan halibut Swiss chard Cress "What a Nut" Salad	Chipotle Flank Steak Asparagus Grilled Romaine Salad	Chicken Under a Brick Zucchini Tricolore Salad

SAMPLE MEAL PLAN, PHASE 1: Week 3 and Beyond

	MONDAY	TUESDAY	WEDNESDAY	THURSDAY	FRIDAY	SATURDAY	SUNDAY
BREAKFAST	**Spinach and Cottage Cheese Omelet** Apple	**Nutty Smoothie** (with berries)	2 scrambled eggs with low-fat ricotta cheese	**Yogurt with Nutty Granola** Blueberries	**CremeSicle**	**Frittata** with Mushrooms ½ grapefruit	High-protein/low-carb bar
SNACK	¼ cup raw nuts/seeds	¼ cup raw nuts/seeds	¼ cup raw nuts/seeds	¼ cup raw nuts/seeds	¼ cup raw nuts/seeds	¼ cup raw nuts/seeds	¼ cup raw nuts/seeds
LUNCH	**Red, White, and Green Salad** with cubed chicken breast	Arugula salad with sliced flank steak leftovers	Hamburger patty wrapped in lettuce leaf Cabbage and carrot slaw	**With Apologies to Joe's Special** Cucumber spears and radishes	**Spa-Ghetti and Bean Soup** Romaine salad	Cobb salad	**Sugar Snap Pea and Mint Soup** with cubed Canadian bacon
SNACK	¼ cup raw nuts/seeds	¼ cup raw nuts/seeds	¼ cup raw nuts/seeds	¼ cup raw nuts/seeds	¼ cup raw nuts/seeds	¼ cup raw nuts/seeds	¼ cup raw nuts/seeds
DINNER	**Chipotle Flank Steak** over stir-fried greens and onions Spinach salad	**Mustard Pork Tenderloin** Braised escarole Caesar salad (no croutons)	**Simply Grilled Alaskan Salmon** Grilled asparagus Romaine and tomato salad	Turkey breast Sautéed spinach with pine nuts Radiccio and endive salad	Alaskan halibut **Roasted Cauliflower with Sage** Watercress salad	Beef Filet Steamed artichoke Boston Lettuce and arugula salad	Barbecued chicken breast Steamed broccoli rabe Tossed green salad

SAMPLE MEAL PLAN, PHASE 2: Week 1 and Beyond

	MONDAY	TUESDAY	WEDNESDAY	THURSDAY	FRIDAY	SATURDAY	SUNDAY
BREAKFAST	Mixed pepper omelet Half grapefruit	Berry-Green Smoothie	2 scrambled eggs with salsa and avocado	Yogurt with Nutty Granola Raspberries	Berry-Green Smoothie	With Apologies to Joe's Special	Evo-Pumpkin Blueberry Pancakes
SNACK	¼ cup raw nuts/seeds	¼ cup raw nuts/seeds	¼ cup raw nuts/seeds	¼ cup raw nuts/seeds	¼ cup raw nuts/seeds	¼ cup raw nuts/seeds	¼ cup raw nuts/seeds
LUNCH	Salad of romaine, chopped broccoli, cherry tomatoes, and tuna	Caesar salad (no croutons) topped with 1/2 chicken breast	Boca Burger or other high-protein veggie burger wrapped in lettuce leaf Raw vegetables	"Dream" of Broccoli Soup Mixed green salad	Tricolore Crab Salad	Red, White, and Green Salad with sardines	Cress "What a Nut" Salad
SNACK	¼ cup raw nuts/seeds	¼ cup raw nuts/seeds	¼ cup raw nuts/seeds	¼ cup raw nuts/seeds	¼ cup raw nuts/seeds	¼ cup raw nuts/seeds	¼ cup raw nuts/seeds
DINNER	Coffee Bean and Peppercorn Crusted Flank Steak Stir-fried greens and onions Spinach salad	2 burritos (low-carb tortilla, Lighthouse taco "meat," romaine, avocado, and fresh salsa)	Not-Bad Pad Thai with shrimp, served over blanched mung beans Arugula salad	Alaskan halibut Roasted asparagus Tossed green salad	Fried Chicken with Almond "Breading" Steamed bok choy Watercress salad	Angelic Jungle Princess with Chicken, over shredded cabbage Roasted Cauliflower with Sage	Tempeh and Black Soybean Quesadilla Casserole Grilled Green Beans

SAMPLE MEAL PLAN, PHASE 2: Week 2 and Beyond

	MONDAY	TUESDAY	WEDNESDAY	THURSDAY	FRIDAY	SATURDAY	SUNDAY
BREAKFAST	**Spinach and Cottage Cheese Omelet** Apple	**Berry-Green Smoothie**	2 scrambled eggs with salsa and avocado	**Dr. G's Koffi-Mocha Freeze**	**Berry-Green Smoothie**	**Mushroom Omelet** served over wilted arugula	**Frittata** with asparagus
SNACK	¼ cup raw nuts/seeds	¼ cup raw nuts/seeds	¼ cup raw nuts/seeds	¼ cup raw nuts/seeds	¼ cup raw nuts/seeds	¼ cup raw nuts/seeds	¼ cup raw nuts/seeds
LUNCH	Chicken breast and slaw wrapped in lettuce leaf	**Seed-Sar Salad**, with tuna	Boca Burger or other high-protein veggie burger, wrapped in lettuce leaf Coleslaw	**"Dream" of Broccoli Soup** Raw vegetables	**Spa-Ghetti and Bean Soup** Boston lettuce salad	**Greens Squared Soup**, with chopped ham Raw vegetables	Arugula salad with sardines Raw vegetables
SNACK	¼ cup raw nuts/seeds	¼ cup raw nuts/seeds	¼ cup raw nuts/seeds	¼ cup raw nuts/seeds	¼ cup raw nuts/seeds	¼ cup raw nuts/seeds	¼ cup raw nuts/seeds
DINNER	**Cajun Blackened Tempeh**, over stir-fried chard, scallions, and garlic **Tricolore Salad**	**Chili-Chicken Double Green Noodles Balsamic Asparagus** Mixed green salad with avocado	Alaskan salmon **Roasted Cauliflower with Sage** Asparagus salad, with sesame seeds and sesame oil and vinegar dressing	**Commander Gundry's Pecan Fish**, served over sautéed kale and onions **Grilled Romaine Salad**	**Stir-fried Tempeh with Green Beans and Basil** Spinach and red onion salad	Beef carpaccio with arugua and lemon juice and olive oil dressing Steamed artichoke **Caprese Salad**	**Dr. G.'s South by Southwest Chicken Sag Paneer** Chopped salad of green peppers, cucumber, tomato, and scallions

SAMPLE MEAL PLAN, PHASE 3: Longevity

	MONDAY	TUESDAY	WEDNESDAY	THURSDAY	FRIDAY	SATURDAY	SUNDAY
BREAKFAST	Poached eggs Kiwifruit (with skin)	Nuts and/or seeds Apple	**Berry-Green Smoothie**	Nut and flaxseed granola with berries	**Berry-Green Smoothie**	**Nutty Smoothie**	**Frittata** with chard
SNACK	⅛ cup raw nuts/seeds	⅛ cup raw nuts/seeds	⅛ cup raw nuts/seeds	⅛ cup raw nuts/seeds	⅛ cup raw nuts/seeds	⅛ cup raw nuts/seeds	⅛ cup raw nuts/seeds
LUNCH	Salad of romaine lettuce, chopped broccoli, cherry tomatoes, tuna	Caesar salad (no croutons) with sliced avocado and/or anchovies	Boca Burger, wrapped in lettuce leaf Raw vegetables served with tahini	**Evo-Pizza**, with raw mushrooms and vegetables	**Evo-Quesadilla** Arugula and red cabbage salad with raw vegetables	**Seed-Sar Salad,** with sliced avocado or Boca Burger	Cobb salad
SNACK	⅛ cup raw nuts/seeds	⅛ cup raw nuts/seeds	⅛ cup raw nuts/seeds	⅛ cup raw nuts/seeds	⅛ cup raw nuts/seeds	⅛ cup raw nuts/seeds	⅛ cup raw nuts/seeds
DINNER	**Raw "Not-Bad" Pad Thai** Spinach and endive salad	**Angelic Jungle Princess** (meatless), with asparagus and mushrooms Bibb lettuce, tomato, and red onion salad	Shaved raw artichoke with Parmesan cheese shavings **Caprese Salad Balsamic Asparagus**	**Sugar Snap Pea and Mint Soup,** topped with chunks of tempeh or Boca Burger Romaine and watercress salad	**Simply Grilled Alaskan Salmon,** served over **Tricolore Salad**	**Nutty-Broccoli Spa-Ghetti** Cabbage, carrot, and daikon slaw	**Chicken Under a Brick,** served over romaine lettuce

DIET EVOLUTION RECIPES

In the recipe section that follows, I've provided easy-to-prepare meals that serve as a guide for the types and quantities of foods that will help you achieve your personal Diet Evolution. In almost every case you'll find options for converting the basic Phase 1 recipe to a Phase 2 and, if appropriate, a Phase 3 version. In some cases, you may prefer not to convert a recipe. If so, no sweat! If you're fond of a Phase 1 recipe but have advanced to Phase 3, simply continue to follow the original recipe, but have it less frequently or reduce the animal protein content.

I've found that my patients who have succeeded on the program—as well as most people in general—have a repertoire of about five or six basic main meals they eat on a rotating basis. Examine your own dining patterns and you'll probably find you do the same. Feel free to substitute fresh ingredients depending on what's available at your store or farmer's market.

HOW TO EVOLVE YOUR SHOPPING STYLE

Although most of my recipes call for familiar ingredients, some ingredients such as agave syrup or whey protein powder may be new to you—and you may not know where to find them. Others, such as tomato sauce with no added sugar or non-Dutch-processed cocoa powder, differ in important ways from those you may be currently using. Once you try some of these ingredients and realize how they increase your options and ability to follow Diet Evolution, I think you will find them as essential as I do. Here are some of my favorites:

Agave syrup Also called agave nectar, this low-glycemic sweetener made from the cactus of the same name is available in natural foods stores and increasingly in grocery stores. The syrup is usually fermented to produce tequila.

Almond meal Ground almonds, available in natural foods stores. (Almond flour is even more finely ground.)

Almond milk Use only unsweetened products. Don't be fooled by terms like "lite" and "low-fat."

Cajun seasoning Sold as Emeril's, Paul Prudhomme's, and Tony Chachere's, this herb/spice mix adds heat to meat, poultry, and fish. Also called Creole seasoning.

Cocoa nibs These raw, shelled, chopped cocoa beans are available at Whole Foods and most natural foods stores.

Cocoa powder Not to be confused with cocoa powder mix, which is sweetened. Use only non-Dutch-processed (non-alkali) products. My favorites are Dagoba and Scharffen Berger.

Green powders These are great ways to get a lot of servings of vegetables into a shake or smoothie. Berry Green, Dr. Schulte's, Very Green, and Miracle Greens or Reds can all be found in natural foods stores.

Hemp protein powder Made from hemp seeds (a cousin of marijuana, and no, you will not feel high eating it), this powder contains the essential amino acids and has the benefits of whey protein powder without the downsides. (Many whey powders contain sugar or artificial sweeteners.) The seeds are high in fat, albeit mostly omega-3s.

High-fiber pasta See page 216 for information.

Mozzarella Use only fresh mozzarella, which comes in baseball-size balls (as well as smaller ones) packed in water.

Protein powder A combination of whey, soy, and rice proteins, made by Rainbow Light in chocolate and vanilla flavors.

Rice protein powder Like whey protein powder, it contains the essential amino acids but is low in taurine, as are most plant-based proteins. Vegans who wish to avoid whey products can use it.

Salsa Read the label to make sure that there is no sugar or corn syrup. I prefer the fresh varieties available now in most supermarkets to the canned or bottled types. You can use tomatillo as well as tomato salsa.

Shirataki tofu noodles See page 216 for information on this stand-in for high-carb pasta.

Soy flour Finely ground soybeans, useful for "breading" and as a substitute for wheat flour in certain recipes; available in natural foods stores.

Soymilk Use only unsweetened products such as those made by WestSoy, Pacific, and Trader Joe's. Don't be fooled by terms like "lite" or "low fat."

Soy protein powder A vegetarian alternative to protein or whey powders, soy protein powder can be used in shakes and smoothies. Be sure to select a product without any added sugars, such as Twinlab's VegeFuel.

Stevia Unlike artificial no-calorie sweeteners, stevia is a natural product, an herb about 300 times sweeter than sugar. Although it gets bitter when exposed to heat, stevia can be used to sweeten smoothies and other uncooked foods. You'll find it in packets, boxes, or convenient sprinkle-top containers in any natural foods store and in some well-stocked grocery stores. A little goes a long way.

Tempeh Fermented soybeans formed into high-protein blocks, tempeh is available refrigerated or frozen in natural foods stores.

Tortillas The only acceptable ones are low-net-carb, high-fiber brands such as La Tortilla Factory or Santa Fe Tortilla Factory. They are sometimes marketed as South Beach tortillas.

Veggie burgers My favorite is Boca Burgers. You can also use veggie crumbles; LightHouse ground beef style vegemeat is a good choice. Avoid Garden Burgers, which are mostly grain based.

Whey protein powder A by-product of making butter, it comes plain or flavored. Read the labels carefully. Many whey powders are loaded with sugars or artificial sweeteners.

Yacón Native to the Andes and related to sunflowers and Jerusalem artichokes, the yacón plant has edible roots full of indigestible fruit sugars, meaning it adds sweetness without impacting blood sugar levels. Yacón can be found in natural foods stores as a powder or syrup. Use it as a substitute for agave syrup in recipes or sweetening beverages.

Yogurt Use only unsweetened, unflavored yogurt, preferably organic and from grass-fed cows. Straus Family Creamery, Natural by Nature, and Organic Valley are good organic choices, as is Greek yogurt such as Fage.

TOOLS FOR SUCCESS

The chances are that you already have everything you need in your kitchen to get cooking the healthy Diet Evolution way. Here is the checklist of the tools you'll need:

Blender or food processor One or both will help you make smoothies, soups, and a host of other goodies. I have found the powerful and indestructible Vita-Mix brand blender increasingly indispensable as my diet has evolved.

Coffee grinder or spice mill

Garlic press

Grill pan and/or grill, or George Foreman-type indoor griller

Measuring cup

Salad spinner

Snack-size plastic bags Measure nuts or seeds in the palm of your hand and store them in individual portions.

Vegetable peeler Your portions of cheese will seem larger if you shave them into thin slices.

Wok or deep frying pan

BREAKFAST

The number-one question I get from my Club members when I tell them about Diet Evolution and the mischief lurking in breads and cereals is, "What can I eat for breakfast?" I tell them that until 1906, no one ate cereal for breakfast because it didn't exist. My English friends remind me that cold cereal didn't even make it to the British Isles until the Yanks brought it with them during World War II. Since we've been eating this new stuff, is anybody getting healthier? Almost all the patients on whom I've performed heart surgery used to have cereal for breakfast, 80 percent of them with a "healthy" banana and "healthy" skim milk. Instead, enjoy eggs—they won't kill you, unlike cereal and other grain-based foods—and open your mind to other choices, such as protein smoothies or even soup or leftovers of last night's dinner.

EGG DISHES

Eggs fortified with omega-3 fatty acids form one foundation of Phase 1 breakfast choices. Despite what you've been told, consuming up to four eggs a day has no major impact on most people's cholesterol. If convenience is an issue, use Eggbeaters or other liquid egg products, but your ancestors did well with the real thing and so can you. Poached, soft-boiled, or sunny-side-up eggs minimize the impact of cholesterol—whipping exposes it to air, which oxidizes it. Despite this, omelets provide a perfect medium with which to introduce greens and other vegetables into a rapidly cooked, easily consumed form.

Basic Omelet

PHASE 1, SERVES 1

2 or 3 omega-3 eggs

¼ cup unsweetened soymilk

Sea salt and cracked black peppercorns to taste

1 teaspoon extra-virgin olive or canola oil

Combine the eggs, soymilk, salt, and pepper in a small, deep bowl or blender. Whisk or blend until almost frothy.

Place a nonstick omelet pan or small frying pan over medium-high heat. Add the oil and spread around pan. Add the egg mixture to the pan and gently push the edges of the cooking eggs toward the middle of pan, tilting the handle to redistribute uncooked areas.

Turn the heat to low. Add additional salt or pepper, if desired, then flip the omelet with a nonstick spatula. If adding vegetables, place on one half of the omelet. Fold the other side of the omelet over and cook briefly to finish. Turn out onto a plate and enjoy.

Spinach and Cottage Cheese Omelet

PHASES 1 AND 2, SERVES 1

1 teaspoon extra-virgin olive or canola oil

1 cup frozen or fresh chopped spinach

¼ cup chopped red or yellow onion, or
    1 tablespoon dried minced onion

1 teaspoon dried thyme or sage

Sea salt and cracked black peppercorns to taste

Ingredients for Basic Omelet (opposite)

½ cup low-fat cottage cheese

Shaved Parmesan cheese, for topping (optional)

Heat the oil in an omelet pan over medium-high heat.

Add the spinach, onion, thyme or sage, salt, and pepper and sauté until water from the spinach is evaporated and onion is translucent. Add the beaten eggs and prepare like Basic Omelet. After flipping the egg mixture, top one half of the omelet with the cottage cheese and gently fold the other half of the omelet over it. Cook until set. If desired, garnish with shaved Parmesan cheese. Turn out onto a plate and enjoy.

VARIATIONS

Spicy Indian Omelet: Stir 1/4 teaspoon curry powder or turmeric into the cottage cheese before adding it to the egg mixture.

Spicy Italian Omelet: After the first two weeks on Diet Evolution, add 1/4 cup chopped and seeded fresh tomato and 1 tablespoon chopped fresh basil (or 1/2 teaspoon dried) to the cottage cheese before adding it to the egg mixture.

Or substitute 2 tablespoons of chopped sun-dried tomato or 1 teaspoon tomato paste for the fresh tomato.

Spicy Mexican Omelet: After the first two weeks on Diet Evolution, add 1/2 teaspoon dried oregano and 1/4 teaspoon chili powder to the cottage cheese before adding it to the egg mixture. Top the cooked omelet with 1/4 cup fresh or sugar-free bottled salsa, or a dollop of guacamole.

Mushroom Omelet

PHASES 1–2, SERVES 1

1 teaspoon extra-virgin olive or canola oil

2 cups sliced or chopped mushrooms
   (button, cremini, shiitake, or other fresh
   or reconstituted mushrooms)

¼ cup chopped red or yellow onion, or
   1 tablespoon dried onion

1 teaspoon dried thyme or sage

Sea salt and cracked black peppercorns to taste

Ingredients for Basic Omelet (page 184)

½ cup low-fat cottage cheese

Shaved Parmesan cheese, for topping (optional)

Heat the oil in an omelet pan over medium-high heat. Add the mushrooms, onion, thyme or sage, salt, and pepper. Sauté until the liquid from the mushrooms evaporates and the onion is translucent.

Add the beaten eggs and prepare like Basic Omelet. After flipping the egg mixture, add the cottage cheese to one side of the omelet and fold the other half over it. Cook a minute more or until set. If desired, garnish with shaved Parmesan cheese. Turn out onto a plate and enjoy.

PHASES 2–3

As you progress through Diet Evolution, you'll adapt the omelet recipes above by:

- Decreasing the number of eggs from three to two and finally to one, while increasing the amount of vegetables

- Reducing or eliminating the cottage cheese

- Alternatively, holding at two eggs, but gradually increasing the vegetables from 1 to 2 cups and mushrooms to 3 cups. My favorite combination is sliced asparagus spears with mushrooms. For a real taste treat, "grill" the asparagus in the frying pan before adding the eggs.

Remember, *the more you eat green, the more you become lean!*

FRITTATAS

Frittatas are a marvelous way of varying any omelet recipe. Adjust the oven rack to accommodate your omelet pan (make sure it has a metal handle without any plastic on it). Set the oven on broil to preheat. Proceed with any of the above omelets, but instead of flipping the omelet, remove the pan from the heat; the egg mixture will still be runny. Sprinkle the flavoring mixture over the eggs, shave a few slices of Parmesan cheese over the top, if desired, and put the pan under the broiler. Watch closely and broil until the top is bubbly and just starting to brown. (Each oven is different, but usually the top rack is best for broiling.) Don't overcook! Remove and slide onto a serving plate.

Evo-Pumpkin Blueberry Pancakes

PHASES 1 (AFTER THE FIRST TWO WEEKS)–3, SERVES 2

Here's some comfort food I cooked up for friends one chilly morning that won't turn on the "Store Fat for Winter" program. In a pinch, you can use old-fashioned rolled oats. For an occasional treat, serve with sugar-free pancake syrup, but avoid the syrup in Phases 2 and 3.

⅓ cup steel-cut oats (Irish or Scottish)

1 tablespoon oat bran (optional)

½ cup egg whites (preferred), or
    4 large omega-3 eggs

3 tablespoons soy flour

1 tablespoon fat-free or low-fat cottage cheese

2 tablespoons canned pumpkin (not pumpkin
    pie mix), preferably organic

1 teaspoon pumpkin pie spice

¼ teaspoon vanilla extract

¼ teaspoon baking powder

3 tablespoons unsweetened plain or vanilla soymilk

½ packet or ¼ teaspoon stevia or other no-calorie
    sweetener, or 1 teaspoon agave syrup, or to taste

¼ cup fresh or frozen blueberries, plus additional
    for serving if desired

Preheat the oven to 200°F.

Combine all the ingredients except the blueberries in a blender or use a hand mixer until blended. Fold in the 1/4 cup blueberries with a spoon or spatula.

Spray nonstick canola oil spray on a griddle pan or other flat-bottomed nonstick fry pan and heat over medium-high heat.

Drop a generous ladle of batter onto the griddle and cook until an occasional bubble appears, about 1 minute. Flip and cook until light brown and puffy. Place on a plate and set into the oven to stay warm while you finish cooking remaining pancakes.

Serve with fresh blueberries, if desired.

PHASES 2–3

Add 1/2 cup grated zucchini or chopped spinach (don't cringe, you won't even know it's there). Hold the soymilk, as the vegetables provide liquid.

Yogurt with Nutty Granola

PHASE 1, SERVES 1

Do without the berries or cherries for the first two weeks of Phase 1. Although you may want to use sweetener at first, use progressively less as your diet and taste buds evolve. Be sure to grind the flaxseeds in a spice mill or coffee grinder so you can digest them.

> 1 cup regular or low-fat organic
>   plain yogurt
>
> ½ cup fresh or frozen blueberries,
>   raspberries, mixed berries, or
>   pitted dark cherries
>
> ½ packet or ¼ teaspoon (or less) stevia
>   or other no-calorie sweetener
>
> 2 tablespoons coarsely chopped raw
>   mixed nuts, or one type
>
> 2 tablespoons freshly ground flaxseeds

Stir together the yogurt and berries with as little sweetener as possible.
Stir together the nuts and flaxseeds
Top the yogurt-berry mixture with the nut-flaxseed mixture. Stir together or keep the nuts as a "granola" topping.

MIX YOUR OWN YOGURT

Most prepared yogurts (even the low-fat, artificially sweetened types) are so sweet that they have no place in any phase of Diet Evolution. But you can get the benefits of the friendly bacteria in yogurt by using plain yogurt and mixing it with berries and nuts or seeds.

PHASE 2

Gradually reduce the amount of yogurt and sweetener and increase the amount of berries.

PHASE 3

Eliminate the yogurt and sweetener and have yourself a bowl of nuts and berries.

DIET EVOLUTION SMOOTHIES

Smoothies are one of the best and quickest ways to introduce lots of easily digested protein and gradually "hide" green foods in a palatable, portable meal.

Nutty Smoothie

PHASE 1, SERVES 1

¼ cup raw almonds, pistachios, walnuts, or pumpkin seeds

1 cup unsweetened vanilla soy or almond milk

2 scoops vanilla or chocolate whey protein powder

1½ cups ice cubes

½ packet or ¼ teaspoon stevia or other no-calorie sweetener, or to taste

¼ teaspoon nutmeg, cinnamon, cardamom, or pumpkin pie spice

Pulse all the ingredients in a blender at high speed until smooth. If too thick or the motor bogs down, add a small amount of water until desired consistency is reached. Pour into a glass and enjoy.

VARIATIONS

Add one or more of these optional extras:

· 1/2 to 1 cup fresh or frozen berries (after first two weeks)

· 1 tablespoon non-Dutch-processed (non-alkali) powdered cocoa (not cocoa powder mix)

· 1/2 frozen underripe (greenish-yellow) peeled banana (after first two weeks)

· 1/2 to 1 measuring scoop green powder (gradually adjust amount upward as you get used to the taste)

· 1 tablespoon ground flaxseeds

CremeSicle

PHASE 1, SERVES 1

Add one or two peeled seedless tangerines, such as Satsuma or Clementine, or one peeled navel orange with a 1-inch piece of peel to the Nutty Smoothie (page 193) mixture and eliminate the spices.

PHASE 2

Gradually diminish the sweetener. Reduce the protein powder to 1 scoop; or replace the whey protein powder with rice or hemp protein powder. Gradually increase the berries to 1 cup.

VARIATION

Berry-Green Smoothie: Gradually add any "leaves" you have in the refrigerator to the basic mixture: start with lettuce, then spinach, broccoli florets, etc. Gradually work your way up to 1 or 2 cups per smoothie.

PHASE 3

Decrease nuts to 1/8 cup. Throw in any veggies you want.

ALL SMOOTHIES ARE NOT CREATED EQUAL

Warning: Smoothies and Frappuccinos sold at the mall or your favorite coffee shop are high-calorie sugar bombs.

Dr. G's Koffi-Mocha Freeze

PHASES 1–2, SERVES 1

Working with the fantastic owners of Koffi, the hippest and busiest coffee house in Palm Springs, I came up with this drink, which is not only delicious but also good for you.

1½ cups ice cubes

1 cup unsweetened soy or almond milk

2 scoops chocolate whey protein powder or
    Rainbow Light Chocolate Protein Powder

1 tablespoon non-Dutch-processed cocoa powder

1 sprig fresh mint, or ⅛ teaspoon mint extract

1 tablespoon instant coffee granules, or
    1 or 2 shots of cooled espresso

½ packet or ¼ teaspoon stevia or other no-calorie
    sweetener, or to taste

1 teaspoon agave syrup (optional)

Pulse all the ingredients in a blender on high until smooth. If the freeze is too thick, thin with small amounts of water while blending. Pour into a glass and drink immediately.

VARIATION

Dr. G's Coconutty Koffi-Mocha Freeze: Eliminate the mint and replace with 1/4 cup unsweetened coconut flakes and/or 1/4 teaspoon coconut extract and blend.

SNACK

To avoid overindulging in this treat, portion out 1/4 cup (1/8 cup for Phase 3) servings into individual plastic bags. All nuts should be unsalted.

MAKES 15 CUPS

1 pound raw shelled walnuts

1 pound raw shelled pistachios

1 pound raw or roasted macadamia nuts

1 pound dry-roasted peanuts

1 pound raw pumpkin seeds or pepitas

Mix ingredients in a large mixing bowl, then place individual portions in the refrigerator or freezer.

LUNCH

The midday meal is a real opportunity to bring your genes in contact with greens, without behaving like a rabbit. Remember, it's a gorilla, not a rabbit, that you're emulating! The more greens in your system, the better you'll feel, I promise. In Phase 1, the object of the game is to include an animal-based source of protein (if you're not a vegetarian or vegan) as part of your salad. As you move to Phases 2 and 3, the proportions of animal-based protein to vegetables will shift significantly in favor of vegetables. Your simplest option is a salad topped with grilled chicken or shrimp.

All these salads are equally suitable as a side dish or first course at dinner. Most of the salad recipes can easily be adapted to serve as a complete lunch or dinner. (Look for "Make It a Meal" in several of the salad and other recipes that follow.) Otherwise, serve them with a burger or other protein source or with a bowl of soup.

Many of my recipes include packaged greens, which are great time-savers. While prewashed salad greens have been washed three times before being packaged, you may feel more comfortable giving them one more washing and then spinning them dry in a salad spinner.

Seed-Sar Salad

PHASES 1–2, SERVES 2

One of my most requested recipes from my patients, and local chefs, the "Cae-sar" dressing for this salad is equally good as a dip for veggies or drizzled over grilled or steamed asparagus. Use pre-washed romaine lettuce, as I do, and you'll make this salad more often.

FOR THE DRESSING

½ cup raw pumpkin seeds or pepitas

1 garlic clove, crushed,
    or ⅛ to ¼ teaspoon garlic powder

¼ teaspoon each sea salt and cracked black
    peppercorns, plus additional pepper
    for serving

Juice of ½ lemon, preferably Meyer

1 teaspoon Dijon mustard

4 tablespoons extra-virgin olive oil

FOR THE SALAD

1 (5-ounce) package pre-washed romaine leaves,
    or 1 head romaine, leaves washed and torn

Choice of: 2 grilled chicken breast halves or
    boneless thighs (available in most grocery stores);
    1 cup cooked cocktail shrimp; 1 (6-ounce) can
    albacore tuna in water, drained; or 1 (5- or
    6-ounce) can crab meat

Shaved Parmigiano-Reggiano cheese, for garnish

MAKE THE DRESSING:

In the bowl of a small food processor, pulse the pumpkin seeds until finely ground but not a paste, or pulse in small batches in a spice or coffee grinder.

If using a food processor, add the garlic, sea salt, and pepper, and pulse until the garlic is well mixed with the seeds. Squeeze in the lemon juice, add the mustard and oil, and blend until creamy. Check and add more salt if desired. Dressing should be thick, but if it is too thick, add more oil or lemon juice. (The dressing can be made and stored in the refrigerator for several days. If it separates, simply whisk briefly before serving.)

MAKE THE SALAD:

Place the salad greens in a large mixing bowl, pour on the dressing, and mix well until the leaves are coated. Place on two plates and top with your choice of protein. Shave a few strips of Parmigiano-Reggiano over the top and sprinkle with a grinding or two of fresh pepper.

PHASE 3

· Eliminate the animal-based protein and cheese.

· Peel, seed, and slice one ripe Hass avocado over the salad in lieu of meat and cheese.

· Replace the animal-based protein and cheese with Cajun Blackened Tempeh (page 200).

Cajun Blackened Tempeh

PHASES 1–3, SERVES 2

Here's my favorite way to serve tempeh, which I have modified from a recipe of Native Foods, a vegan restaurant in Palm Springs and Los Angeles. Add extra Cajun seasoning if you want more heat!

1 (11-ounce) block tempeh, thawed (if frozen) and sliced into ¼-inch-thick strips

3 tablespoons light soy sauce or tamari

1 tablespoon Cajun seasoning, such as Emeril's

2 tablespoons extra-virgin olive oil

Toss the tempeh strips in the soy sauce, then sprinkle with or roll in Cajun seasoning.

Heat the oil in a nonstick or cast iron frying pan over medium-high heat. Add the tempeh and brown for 2 minutes; turn and brown on the other side for 2 minutes more.

Caprese Salad

PHASES 1–2, SERVES 2

This combination of fresh basil, tomatoes, and mozzarella (particularly if you can find water buffalo mozzarella), topped with extra-virgin olive oil and a drizzle of balsamic vinegar, can make a meal. Even though aged balsamic vinegar de Modena is expensive, a little goes a long way.

2 large ripe tomatoes, such as beefsteak,
     heirloom, or Roma

8 ounces fresh mozzarella

1 cup, packed, fresh basil leaves

4 tablespoons extra-virgin olive oil

1 tablespoon balsamic vinegar (optional)

Slice the tomatoes into approximately 1/4-inch-thick slices and arrange on two plates. Slice the mozzarella to the same thickness and place on top of the tomato slices.

Finely chop the basil with a very sharp knife and combine in small bowl with the oil. Spoon the dressing over the tomato and cheese slices. Serve.

NOTE: For the traditional Italian version, stop here. I've never been served balsamic vinegar on Caprese salad in either Italy or France. Never. But, it's sooooo much better if you do drizzle a little over the salad.

Red, White, and Green Salad

PHASE 1, SERVES 2

1 (8- to 12-ounce) package cherry or grape tomatoes

8 ounces fresh marble-size mozzarella balls
(*boccotini*) or larger balls cut into 1-inch cubes

½ (5-ounce) bag pre-washed arugula, mâche,
or mixed baby greens

1 cup, packed, fresh basil leaves

6 tablespoons extra-virgin olive oil

2 tablespoons balsamic vinegar

Sea salt and cracked black peppercorns to taste

Toss together the tomatoes, mozzarella, and greens in a medium serving bowl.

Finely chop the basil and combine with the oil, vinegar, salt, and pepper. Whisk while pouring over salad.

Toss again, place on individual serving plates, and top with a grinding of fresh pepper.

VARIATIONS

Make It a Meal: Those with larger appetites may want to try these Phase 1 variations, which are also suitable for Caprese Salad.

· Add 1 cup cubed grilled chicken breast.

· Add 1 (6-ounce) can albacore tuna packed in water, drained.

· Add 1 (3-ounce) can sardines in oil, drained.

PHASE 2

Substitute one-half sliced avocado for 4 ounces of mozzarella cheese. This works equally well for Caprese Salad.

PHASES 2–3

Substitute a whole sliced avocado for all the cheese.

Cress "What a Nut" Salad

PHASE 1, SERVES 2

While I eat romaine lettuce in my salads daily and never tire of it, this recipe should introduce you to some other flavors and textures so as to avoid any salad monotony. Look for washed watercress in a sealed bag with roots still attached.

DRESSING

4 tablespoons extra-virgin olive oil or walnut oil

2 tablespoons apple cider vinegar or lemon juice

½ teaspoon Dijon mustard

Sea salt and pepper to taste

¼ packet or ¼ teaspoon stevia or other no-calorie
     sweetener (optional)

SALAD

1 bunch watercress

1 (8-ounce) package mung bean sprouts, or
     2 cups loose sprouts, rinsed and drained

½ cup walnut or pecan halves

½ ripe avocado, sliced

¼ cup crumbled Gorgonzola, feta, or bleu cheese,
     or coarsely grated Asiago

Whisk or shake together the dressing ingredients. Chill slightly, if desired. Gently mix the watercress and sprouts; toss with the dressing.

Place equal portions of greens on two plates, then distribute half the nuts, avocado slices, and cheese on top of each and serve.

VARIATIONS

- Eliminate the bean sprouts and substitute a head of butter lettuce, rinsed and torn into bite-size pieces. Mix the ingredients and top with 1/4 cup pomegranate seeds.

- After the first two weeks, top salad with a crisp pear or Asian pear, cored and thinly sliced.

- Serve salad with Sundance Lavender Dressing (page 206).

Sundance Lavender Dressing

The use of lavender transforms a basic salad dressing into one of my favorites. Try this unique dressing on the other salads as well.

8 SERVINGS

¼ cup white wine vinegar

⅓ cup extra-virgin olive oil

⅓ cup canola or walnut oil

½ packet or ¼ teaspoon stevia or other no-calorie sweetener, or 1 teaspoon agave syrup

2 tablespoons finely chopped fresh lavender, or 1 teaspoon dried

Sea salt and cracked black peppercorns to taste

Combine all the ingredients in a deep bowl or small food processor. Refrigerate leftovers and use within a week.

VARIATIONS

Make It a Meal: Top each portion of salad with a palm-size seared Ahi tuna steak. Drizzle the tuna with extra-virgin olive oil, salt, and pepper, and sear over high heat for 1 minute on each side in your grill pan. Slice into 1/4-inch-thick pieces and fan out over salad.

PHASE 2

This is just about perfect as it is. If adding tuna, halve the portion.

PHASE 3

Reduce the amount of oil and eliminate or cut back on the cheese. If expanding to a meal, substitute a few slices of sushi-grade raw tuna, gravlax, or prosciutto.

Tricolore Salad

PHASES 1–3, SERVES 2

Bored with your usual salad makings? Remember, your primitive ancestors dined on hundreds of different bitter greens. Here's an easy way to introduce interesting, colorful greens into your salads because, as you know, more bitter, more better! Also for a difference, serve the salad the authentic way, by placing each green on one-third of a large salad plate, like slices of pizza. Drizzle the dressing over the salad and top with shaved Parmigiano.

SALAD

1 head radicchio

2 Belgium endive

1 head romaine, washed and torn, or ½ (5-ounce) bag pre-washed romaine

Shaved Parmigiano-Reggiano cheese, for garnish

DRESSING

½ cup extra-virgin olive oil

4 tablespoons balsamic vinegar

¼ teaspoon Dijon mustard

Sea salt and cracked black peppercorns to taste

Core radicchio and trim endive ends. Chop radicchio and slice endive crosswise into 1-inch pieces; rinse greens and dry in a salad spinner. Toss together with romaine in a serving bowl.

Whisk the dressing ingredients in a small bowl or pulse in a small food processor or blender. Toss greens with dressing and shave cheese over top. Serve.

Tricolore Cobb Salad: For Phases 1 and 2, toss the basic salad, divide onto two plates, and top each portion with 1 grilled chicken breast, chopped; 1 prepared turkey bacon strip, crumbled; 1/2 Hass avocado, peeled, seeded, and cut into 1/2-inch pieces; and 3 tablespoons crumbled Gorgonzola, bleu cheese, or feta.

Tricolore Crab Salad

PHASES 1–3, SERVES 4

1 tablespoon extra-virgin olive oil

2 garlic cloves, minced

⅛ teaspoon red pepper flakes

½ teaspoon Old Bay seasoning

1 (1-pound) can crab meat (lump and back fin preferred), drained and picked over

1 tablespoon chopped fresh parsley

Juice of ½ lemon

Ingredients for Tricolore Salad (page 207)

Heat the oil over medium-high heat in a wok or deep frying pan. Add the garlic, red pepper flakes, and Old Bay seasoning; cook for 1 to 2 minutes. Add the crab and cook until just heated through, another minute. Stir in the parsley and lemon juice. Place dressed salad on serving plates and top with crab mixture.

Grilled Romaine Salad

PHASE 1, SERVES 2

Most lettuces take to grilling like a duck to water. Once you grill romaine (or radicchio), and top it with warmed dressing and a grilled piece of ham or prosciutto, you'll understand how wonderful a grilled salad can be. If using a charcoal grill, heat the coals ahead so there's a light ash covering when you are ready to grill the lettuce.

¼ cup extra-virgin olive oil, plus more if desired

4 slices turkey bacon, Canadian bacon, deli ham, or prosciutto

1 small red onion, chopped, or 4 tablespoons minced dried onion

¼ cup balsamic vinegar

2 heads romaine, washed and dried

¼ cup crumbled goat cheese or Gorgonzola cheese

Sea salt and cracked black peppercorns

Heat 2 tablespoons oil over medium-high heat in a skillet or wok; add the bacon, and cook until crisp but not burnt, about 2 to 5 minutes. Add the onion and stir-fry until translucent and just starting to brown, 2 to 3 minutes more. Add the vinegar to the pan and reduce the heat to medium. Simmer to reduce the sauce until syrupy. Set aside while you grill the lettuce.

Heat a grill pan over high heat. Leaving the cores intact, cut the lettuce heads lengthwise into quarters. Brush the cut edges with the remaining 2 tablespoons oil. Place on a hot grill, turning after each side browns, 2 to 3 minutes per side. Alternatively, grill under a preheated broiler or fry in a skillet until lightly browned.

Arrange the grilled romaine on two plates and top with the dressing, crumbled cheese, salt, and pepper. If desired, drizzle a little more olive oil over the top.

VARIATION

Grilled Radicchio Salad: Replace the romaine with radicchio. Cut into quarters.

PHASE 2

- Chop a (3- to 4-ounce) can of sardines or 4 anchovy fillets and use in place of bacon or ham.

- Use a (6-ounce) can of water-packed tuna, drained, in place of bacon or ham.

- Use soy-based salami or pepperoni in place of the bacon or ham.

PHASE 3

- Eliminate the bacon or ham and grill a handful of asparagus spears or fresh green beans for 3 to 4 minutes alongside the lettuce; top all with dressing.

- Reduce the amount of cheese or eliminate altogether.

SOUPS

Nothing beats a hearty bowl of soup. It can be your ally in easily getting a large volume of greens into your system—and filling your tummy. A few tweaks to old standards will tickle your taste buds and turn on your longevity genes!

"Dream" of Broccoli Soup

PHASE 1, SERVES 4

Although this soup is tummy-warming in fall and winter, it can also be served chilled in the summer with a dollop of yogurt and garnished with a few sprigs of cilantro or Italian parsley.

1 head broccoli or broccoflower, or about 4 cups
   frozen broccoli florets

1 cup fresh or frozen shelled edamame
   (green soybeans)

1 small red onion, chopped

6 cups low-sodium, low-fat chicken stock or bouillon

½ cup low-fat cottage cheese or ricotta

1 tablespoon chopped fresh mint (or basil,
   if not available)

Sea salt and cracked black peppercorns to taste

Coarsely chop the fresh broccoli or broccoflower. Put the chopped broccoli or frozen florets, edamame, onion, and chicken stock in a large saucepan. Bring to a boil over medium-high heat, then reduce the heat, cover, and simmer for 10 minutes.

Pour the contents of the saucepan into a blender or a food processor and purée until smooth. Add the cheese and mint; blend ingredients, then pour soup back into the saucepan. Check for taste and add salt and pepper. Reheat slightly if necessary before serving.

VARIATIONS

- Add 1 cup chopped Canadian bacon, deli ham, or prosciutto to the soup and simmer for a few minutes before serving.

- Add 1 cup shredded grilled chicken and 1/4 teaspoon Tabasco or other hot sauce, and return to a simmer for a few minutes before serving.

- Use 1/2 bunch broccoli rabe, a more bitter form of broccoli, or 1 bunch kale or collard greens in place of half the broccoli.

PHASES 2–3

You have several options to make this soup suitable for later phases:

- Reduce or eliminate the edamame and/or cheese.

- Reduce or replace chicken stock with vegetable stock or water.

- For a chunkier version, reserve 1 cup cooked broccoli, rather than blending it, then return to soup and reheat.

- After returning soup to saucepan, add 2 to 3 cups baby spinach, chopped Swiss chard, or chopped mustard greens and heat until wilted but still very green.

PHASE 3

Raw Broccoli Soup: Combine the uncooked broccoli, onion, salt, and pepper, and pulse in a blender with 3 cups water until creamy. Adjust to desired thickness, adding more water as necessary.

Sugar Snap Pea and Mint Soup

PHASE 1, SERVES 4

Who doesn't like the comforting warmth of split pea soup? Here's my take on a classic, made "user friendly" for our less active lives. Substitute snow peas if you can't find sugar snap peas and basil if mint isn't available.

> 3 cups stringless sugar snap peas, fresh or frozen, washed
>
> 1 cup fresh or frozen shelled edamame (green soybeans)
>
> 4 cups chicken stock
>
> ½ cup regular or low-fat cottage cheese
>
> 1 tablespoon chopped fresh mint
>
> Sea salt and cracked black peppercorns

Combine the peas, edamame, and stock in a saucepan; bring to a boil over medium-high heat. Reduce the heat and simmer for 10 minutes, or until vegetables are soft. Pour soup into a blender or food processor (a blender will give you better texture) and pulse until creamy. Add the cottage cheese, mint, sea salt, and pepper and blend ingredients again. Pour back into the saucepan to reheat gently, then serve.

VARIATIONS

Greens Squared Soup: After returning the soup to the saucepan, stir in 2 cups baby spinach leaves or mustard greens and heat, stirring, until leaves are wilted and incorporated into the soup.

Green "Bean" Soup: After returning the blended soup to the saucepan, stir in one more cup of cooked edamame and reheat.

Make It a Meal: For Phase 1, after blending the soup, add 1 cup chopped Canadian bacon or smoked ham and heat. Tastes like the classic!

PHASES 2–3

- Substitute water or vegetable stock for chicken stock.

- Substitute 1/2 avocado for cottage cheese.

- Reduce the amount of edamame.

FOOL-DLES

The name of this section refers to the fact that the recipes I've adapted, which use shirataki noodles, will fool you into thinking you're eating pasta. My wife, Penny, who cringed at every phony pasta I tried to pass off on her, can't wait for the next new shirataki noodle dish I've come up with.

Before you dive into these treats, let me tell you my favorite way of getting the noodles ready to use. Do not, I repeat, *do not follow the package directions* or you will likely be disappointed. Instead, rinse the noodles in a colander under cold running water for a minute or two to eliminate the somewhat fishy odor; cooking will remove any residual odor. Then drain and drop the noodles into boiling, salted water in a nonstick wok or large skillet and boil for 2 to 3 minutes. Drain and rinse with cold water again. Return the noodles to the empty wok or skillet and move them around the pan over medium heat to partially dry them out. If you want to enjoy these noodles as "pasta," don't skip this step, which will just take a couple of minutes—otherwise, you will find them slimy. Set dried noodles aside for use in the following recipes.

There is now another low-carb, high-fiber option that passes the Penny test. FiberGourmet has recently come out with a line of pasta that lives up to its name. Each 1-cup serving contains 18 grams of fiber, minimizing its impact on your blood sugar level, and 140 calories. Although better than the 220 calories in a cup of conventional pasta, this is more than shirataki tofu noodles, so watch the serving size. Available in six flavors, including chocolate(!), it may be hard to find in your area but can be ordered from www.fibergourmet.com.

If you use this pasta in the following recipes, be sure to reduce the amount of pasta to 1 cup (cooked) per serving. Figure on a 1/2 cup of dry pasta yielding a cooked 1-cup serving.

Fettuccine Dr. Gfredo

PHASE 1, SERVES 1

I adapted this recipe from Hungry-girl.com, though their Hungry Girlfredo is delicious in its own right. However, the key to enjoying shirataki is in preparing the noodles as I've outlined on page 216.

1 (8-ounce) package tofu shirataki fettuccine noodles, prepared the Dr. G. way

1 wedge Laughing Cow light cheese

2 teaspoons grated Parmigiano-Reggiano cheese

1 teaspoon low-fat sour cream

Cracked black peppercorns to taste

Sea salt (optional)

Keep the prepared noodles warm in a wok or skillet over medium heat. Add both cheeses and the sour cream and stir until all are melted and incorporated. Taste for saltiness—the cheeses may make salt unnecessary—and sprinkle with pepper. Serve.

VARIATIONS

Lemon-Basil Dr. Gfredo: Add 1 teaspoon fresh lemon zest or 1/2 teaspoon dried, 1 teaspoon lemon juice, and 1/2 cup coarsely chopped fresh basil leaves. Heat until basil is wilted.

Chicken Lemon-Basil Dr. Gfredo: Add 1 cup chopped chicken and 1/4 cup pine nuts to the variation above. Heat until warm.

Ham and Arugula Dr. Gfredo: Add 1 cup chopped ham or 4 slices prosciutto, chopped, and 2 cups baby spinach or arugula. Heat until greens are wilted.

PHASE 2

Simply add 2 cups chopped greens of your choice and reduce the amount of meat in the variations.

PHASE 3

As above, but eliminate the meat from the variations.

Not-Bad Pad Thai

PHASES 1–2, SERVES 2

Here's a simple way to cut the calories in Thailand's most famous dish without sacrificing all the spicy goodness.

4 tablespoons extra-virgin olive oil

2 garlic cloves, chopped or crushed

12 medium shrimp (21–25 per pound),
    shelled and deveined

1 omega-3 egg

½ cup basil leaves, chopped

1 (8-ounce) package tofu shirataki fettuccine noodles,
    prepared the Dr. G way (see page 216)

¼ cup lime juice

4 tablespoons crushed dry-roasted low-salt peanuts

1–2 tablespoons fish sauce, soy sauce, or tamari
    (see Note)

1 tablespoon unseasoned, unsweetened rice vinegar

Pinch of stevia or no-calorie sweetener

1 teaspoon paprika

¼ teaspoon red pepper flakes or cayenne pepper,
    more or less to taste

In a large skillet or wok, heat the oil over high heat, but not to the point of smoking. Add the garlic and stir briefly, then add the shrimp and stir for 1 minute. Add the egg and stir for another minute.

Add the basil, noodles, lime juice, peanuts, fish sauce, vinegar, stevia, paprika, and red pepper flakes and stir-fry for about 3 more minutes, or until the shrimp has changed from translucent to opaque. Divide the noodles and shrimp mixture evenly onto two plates.

NOTE: Fish sauce, also known as *nam pla* or *nuoc nam*, is an essential ingredient in Southeast Asian cuisines, but can be an acquired taste. Unless you're accustomed to its flavor, use less at the start. For an acceptable but slightly less authentic taste, soy sauce or tamari can be substituted.

VARIATION

Chili-Chicken Double Green Noodles: Omit the shrimp. Add 1 teaspoon grated fresh ginger or 1/2 teaspoon ground ginger, to the garlic and oil mixture. Slice one free-range chicken breast into 1/2-inch-thick strips, add to pan, and cook for 4 minutes, stirring constantly. Add 1 cup green beans and stir for 2 minutes more before adding remaining ingredients according to directions.

PHASE 2

· Halve the amount of shrimp or chicken.

· *Chili-Tempeh Double Green Noodles:* Substitute tempeh strips for the shrimp and cook as in the Chili-Chicken variation.

PHASE 3

Have it raw, using the recipe on page 221.

Raw "Not-Bad" Pad Thai

PHASES 1–3, SERVES 2

FOR THE SAUCE

1 cup almond butter

½ navel orange or seedless tangerine, peeled
   and seeded

2 tablespoons soy sauce or tamari

1 tablespoon agave syrup

Pinch of stevia or other no-calorie sweetener

2 tablespoons grated fresh ginger

¼ cup water, if needed

4 cups mung bean sprouts, rinsed

1 (8-ounce) package tofu shirataki noodles,
   prepared the Dr. G way (see page 216)

In a food processor or blender, combine the sauce ingredients and pulse until well mixed. If too thick, dilute with small amount of water.

Combine the bean sprouts and shirataki noodles. Divide onto two plates. Cover evenly with the sauce and serve.

SPA-GHETTI RECIPES

Shirataki noodles also come spaghetti style. Follow the same preparation technique as described on page 216, but even when dried in a skillet, the spaghetti strands are still unruly. To tame them, cut the strands into 2- to 3-inch lengths before mixing them with sauce. You can also use FiberGourmet light pasta, but keep portions to 1 cup of cooked noodles. (See page 216.)

Spa-Ghetti and Meat(Less) Balls

PHASE 1, SERVES 2

With a texture more akin to the pasta served at Olive Garden than *al dente*, still this pasta is one I can eat in quantity without feeling guilty or lethargic from the sugar load that regular pasta carries.

2 tablespoons extra-virgin olive oil

1 tablespoon Italian herb seasoning

2 tablespoons minced onion

1 garlic clove, crushed, or ¼ teaspoon garlic powder

1 cup no-sugar-added tomato sauce

1 cup or more fresh basil leaves, baby spinach, or arugula

¼ cup fresh oregano, or 1 teaspoon dried

¼ teaspoon sea salt

¼ teaspoon cracked black peppercorns

⅛ teaspoon red pepper flakes or cayenne pepper (optional)

2 Boca Burgers, broken into pieces, or ½ (14-ounce) package LightHouse ground beef-style vegemeat

1 (8-ounce) package tofu shirataki spaghetti-style noodles, prepared the Dr. G. way (see page 216)

Shaved Parmigiano-Reggiano cheese, for garnish

Combine the oil, Italian seasoning, onion, and garlic in a wok or deep skillet; sauté over medium-high heat until fragrant. Add the tomato sauce, basil, oregano, salt, pepper, red pepper flakes, and Boca Burgers. Simmer until basil is wilted, about 2 minutes.

Add the shirataki spaghetti and stir. Place on two plates and shave cheese over top.

VARIATIONS

· Add 1 cup or more additional spinach, mustard greens, Swiss chard, or arugula.

· Add 1/2 cup turkey pepperoni or salami for the ultimate "meat-lovers" spaghetti.

PHASE 2

· As is, this is a perfect meal when combined with a salad.

· Do without the turkey pepperoni or salami.

PHASE 3

Omit the Boca Burgers and briefly blend three plum tomatoes, 1 cup basil leaves, half an avocado, and the rest of the seasonings in a food processor or blender. Combine the mixture with the prepared spaghetti.

Spa-Ghetti Checca: Dice three Roma or plum tomatoes, chop the basil and two garlic cloves, and mix with 1/4 cup extra-virgin olive oil. Pour over prepared spaghetti.

Nutty-Broccoli Spa-Ghetti

PHASE 1, SERVES 2

Italians are always coming up with ideas to get broccoli into their pasta. Here's a taste-tempting way to get all the benefits of cruciferous vegetables.

> 1 head broccoli or 1 (16-ounce) bag
> pre-washed broccoli florets
>
> ½ cup diced red onion
>
> 3 garlic cloves, peeled
>
> 1 cup chicken broth
>
> 1 (8-ounce) package tofu shirataki spaghetti-style
> noodles, prepared the Dr. G. way (see page 216)
>
> 1 cup water
>
> 2 tablespoons extra-virgin olive oil
>
> ¼ cup raw chopped walnuts
>
> ½ cup feta or Gorgonzola cheese
>
> ⅛ teaspoon red pepper flakes
>
> ¼ teaspoon sea salt
>
> Cracked black peppercorns to taste
>
> Chopped raw walnuts, for garnish

If using a head of broccoli, separate into florets and stems. Cut florets into 1- to 2-inch pieces. If using bagged broccoli florets, cut stems as close as possible to flowers and divide. Cut stems of broccoli into 1-inch pieces and place in a medium saucepan with the onion, garlic, and chicken broth. Bring to a boil, then reduce the heat to low, cover, and simmer for 10 minutes.

Place the noodles, broccoli florets, and water into a second saucepan, bring to a boil, and cook for 4 minutes. Drain in a colander. Return broccoli and noodles to saucepan.

Place the cooked stems, onion, and garlic in a food processor and pulse until smooth. Blend in oil, walnuts, and cheese. Add the red pepper flakes, salt, and pepper and adjust seasonings if necessary. Pour sauce over spaghetti and florets and stir. If desired, top with additional chopped walnuts and serve.

VARIATION

Top cooked florets with 1 cup chopped grilled chicken breast, then pour sauce over dish.

PHASE 2

- These spaghetti recipes are fine for all phases of Diet Evolution, but I like to gradually sneak in more and more greens. Try adding a handful of chopped kale, rapini, watercress, arugula, or any other bitter green to the broccoli stems that become the sauce.

- Add a handful of greens to the broccoli florets.

PHASE 3

- Reduce the amounts of cheese and walnuts.

- Halve the cooking time so as to serve the vegetables half raw.

Spa-Ghetti and Bean Soup

PHASES 1–2, SERVES 4

The classic Tuscan dish *pasta e fagioli* is laden with starch in the form of beans and pasta, but with a few twists it becomes a nutritious, delicious source of protein. Be sure to use *black soybeans*, not black beans, which are available at Whole Foods or any natural foods store.

1 (15-ounce) can diced plum tomatoes

1 (11-ounce) can black soybeans, drained

1 small red onion, chopped, or 4 tablespoons
    dried onion

2 garlic cloves, chopped

2 tablespoons extra-virgin olive oil,
    plus extra for garnish

¼ teaspoon sea salt

¼ teaspoon cracked black peppercorns

1 cup fresh basil leaves, chopped coarsely

2 cups fresh kale, collard greens, turnip greens,
    or chopped spinach, or 1 (10-ounce)
    frozen package

1 tablespoon red wine vinegar (optional)

1 package tofu shirataki spaghetti, prepared
    the Dr. G. way (see page 216), cut into 1- to
    2-inch pieces

Combine the tomatoes, soybeans, onion, garlic, oil, sea salt, and pepper in a medium saucepan. Stir over medium-high heat until the mixture reaches a boil, then reduce heat to simmer, cover, and cook for 15 minutes.

Pour half the mixture (make sure you get some beans) into a blender or a food processor and pulse until coarsely ground and creamy. Return blended mixture to saucepan. Add the basil, kale, vinegar if desired, and noodles. Bring to a boil and cook for 2 minutes more, or until leaves are wilted but still bright green. Serve in bowls. If desired, add a drizzle of olive oil for authenticity.

VARIATIONS

· Add 1 cup chopped ham or cooked chicken breast along with the basil and kale.

· Add 1 cup cooked cocktail shrimp along with the basil and kale.

PHASE 2

This dish is already perfect for this phase. Or substitute 1 (14-ounce) package Light Line ground beef-style vegeburger or 2 Boca Burgers, broken into pieces, instead of the ham, chicken breast, or shrimp.

PHASE 3

Gradually reduce the beans to 1/2 can, and then 1/4 can, while increasing the amount of greens.

Evo-Pizza

PHASE 1, SERVES 2

Far better than what passes for pizza in the United States, this handy "plate" topped with tomatoes, a little bit of cheese, herbs, and various veggies is inspired by pizza I had near Portofino, on the Italian Riviera.

2 low-carb tortillas

2 tablespoons extra-virgin olive oil, plus additional for serving

¼ cup no-sugar-added tomato sauce or chopped fresh plum tomatoes

4 ounces fresh mozzarella (packed in water; buffalo mozzarella preferred), thinly sliced

8 fresh basil leaves, or 1 tablespoon dried

OPTIONAL TOPPINGS

Anchovies, mushrooms, artichoke hearts, prosciutto, Canadian bacon, green and/or red peppers

Sea salt and/or cracked black peppercorns to taste

2 cups arugula (optional)

Red pepper flakes

Preheat the oven to 450°F.

Place the tortillas on a cookie sheet or pizza oven brick. Brush with olive oil. Spread tomato sauce or chopped tomatoes on the tortillas and layer on thin slices of mozzarella. Top cheese with basil leaves and then add one or more additional toppings of your choice, but not arugula. Sprinkle with salt and cracked pepper. Bake approximately 5 minutes, until warm; cheese will not be completely melted and bubbly.

Place pizzas on plates. Top with arugula and sprinkle with red pepper flakes and additional olive oil, if desired.

PHASE 2

- Cut cheese portion in half and double the amount of basil.

- Cut any meat portion in half or substitute crumbled Boca Burgers.

PHASE 3

Place large romaine or butter lettuce leaves on plates, and top with chopped tomatoes, fresh buffalo mozzarella pieces, chopped basil, and any fresh chopped vegetables or mushrooms. Splash on some olive oil, and sprinkle on red pepper flakes, salt, and pepper.

Evo-Quesadilla

PHASES 1 AND 2, SERVES 4

Sometimes the hankering for something between two pieces of bread becomes overwhelming. Keep your sanity by making a quesadilla with low-carb tortillas. Instead of an overload of cheese and meat, this recipe delivers another high dose of greens. Make sure you squeeze the spinach dry or you will be unhappy with me!

1 (8-ounce) bag prewashed spinach, or
    1 (10-ounce) box frozen chopped spinach,
    thawed and squeezed dry

Sea salt to taste

Cracked black peppercorns to taste

16 ounces fresh mozzarella balls (buffalo
    mozzarella preferred), thinly sliced

8 low-carb tortillas

8 ounces Gorgonzola or other creamy
    bleu cheese, crumbled

Cook the fresh spinach over medium-high heat briefly in a covered pan with no extra water until wilted. Drain and squeeze the spinach dry and chop coarsely. Add salt and pepper to cooked or frozen spinach.

Place the mozzarella slices over four tortillas and top with a sprinkling of bleu cheese. Cover with the spinach, evenly dividing among the tortillas. Top each with another tortilla and press together.

Heat a nonstick skillet or flat griddle over medium heat. Place the tortillas in the skillet and cook for 3 minutes, being careful not to burn. Flip and cook until other side is browned and the cheese melts.

Transfer to a cutting board and cut into quarters or halves and serve.

VARIATIONS

Tex-Mex Evo-Quesadilla: Add 1 tablespoon or more of chunky jalapeño salsa to the spinach and substitute sliced avocado for Gorgonzola.

Green Eggs and Ham: Substitute one fried or poached egg for the Gorgonzola. Place one slice of Canadian bacon or thin deli ham or prosciutto over the egg, then top with the second tortilla and cook as above. Serve whole.

PHASE 3

Reduce the amount of cheese by half and double the amount of veggies.

"Rice" and "Beans"

PHASE 1, SERVES 4

Unfortunately, most "comfort food" serves your taste buds but not your genes. In this recipe, high-protein soybeans substitute for beans with a high sugar content, and finely shredded cauliflower stands in for white rice.

1 head cauliflower

¼ cup extra-virgin olive oil

4 chicken or turkey Polish sausages or kielbasa, cut into 1-inch pieces

1 large yellow onion, chopped

1 green bell pepper, seeded and chopped

2 celery stalks, chopped

4 garlic cloves, minced or chopped

2 tablespoons Cajun seasoning, such as Emeril's

¼ teaspoon cayenne pepper

2 teaspoons dried oregano

1 tablespoon dried thyme

1 tablespoon soy flour

1 (16-ounce) bag frozen shelled edamame (green soy beans)

Sea salt and cracked black peppercorns to taste

Tabasco or other hot sauce

Cut the cauliflower into pieces and pass through the shredder blade of a food processor or a cheese shredder to form rice-size pieces. Set aside.

Heat the oil in large nonstick skillet over medium-high heat; add the sausage, onion, bell pepper, celery, garlic, Cajun seasoning, cayenne pepper, oregano, and thyme. Cook, stirring often, until the vegetables have softened and are starting to brown. Stir in the soy flour.

Add the edamame, and cook for 5 more minutes, or until they are fully warmed. Add the cauliflower "rice" and cook, stirring, for 2 minutes to warm through. Add salt and pepper to taste.

Turn off the heat and scoop 1 cup of the mixture into a food processor and pulse until creamy. Return this paste to the skillet and stir. Add a little hot water to adjust the consistency; the dish should be quite thick. Serve in bowls. Add Tabasco or other hot sauce.

PHASE 2

Replace sausages with 4 Boca Burgers or Boca sausages, added along with the edamame. Serve over sautéed spinach.

PHASE 3

Reduce the amount of edamame and Boca "meat" by half, and add 1 (5-ounce) bag of prewashed spinach or 1 (10-ounce) box frozen spinach, thawed. Serve over raw spinach or romaine lettuce.

MEATS, POULTRY, AND SEAFOOD

These animal protein sources are your best friends during Phase 1, as long as you remember that the portion size should be approximately the size and thickness of the palm of your hand. That's right, no fingers, just the palm. You big-handed folks get a larger portion than you little-palm people.

However, as Diet Evolution progresses, keep in mind that these foods should be regarded as a topping for a salad or grilled vegetables, and not as the main dish. With that in mind, the recipes are designed so that meat, poultry, and fish can be sliced into thin strips to layer over your veggies. For this reason, in this section I have generally not provided recipes for Phases 2 and 3. Simply reduce the amount of animal protein with each phase. You'll see that the number of servings listed for each recipe in this section reflects this evolution.

Whenever possible, use grass-fed beef, free-range chicken, and wild-caught fish and shellfish. Don't be fooled by organic labels. In general, it just means the animals were fed organic grain products, not grass. But don't sweat it. Do what you can, with what you have, wherever you are.

Chipotle Flank Steak

PHASES 1–3, SERVES 4 (AND LATER 8)

Flank steak is the leanest, easiest-to-grill beef available. If you're new to Southwestern, Texan, or Thai cooking, you may need to back off on the heat of chipotle chiles by using fewer initially—and removing the seeds—and then adding more as you go.

> 3 tablespoons extra-virgin olive oil
>
> 3 tablespoons lime or lemon juice
>
> 1 tablespoon Dijon mustard
>
> 1 garlic clove, minced
>
> 1 teaspoon ground cumin
>
> 1 tablespoon minced canned chipotle chiles en adobe,
>   or 1 tablespoon pure chile powder
>
> 1 (1¼-pound) flank steak
>
> Sea salt and cracked black peppercorns to taste

Combine all the ingredients except the steak and salt and pepper in a large, heavy-duty resealable plastic bag; slosh to mix well. Add the steak and press out most of the air to maximize contact with marinade. Marinate at room temperature for at least 1 hour or overnight in refrigerator.

Preheat a grill pan or barbecue over high heat.

When grill pan is hot or coals are ready, remove the steak from the marinade and sprinkle with salt and pepper. Grill to desired doneness (it's best medium-rare), about 4 minutes per side. Remove from grill, place on carving board, and let rest for 5 minutes.

Cut steak against the grain into thin diagonal slices. Serve over salad of your choice or grilled asparagus, which you can marinate in the same bag and cook alongside the steak.

Coffee Bean and Peppercorn Crusted Flank Steak

PHASES 1–3, SERVES 4 (AND LATER 8)

Coffee beans and peppercorns are not just good, they're also good for you. In a topping inspired by Miami chef Allen Susser, this recipe takes a humdrum cut of beef to new heights. Be sure to use whole coffee beans; ground coffee just won't cut it.

1 garlic clove, minced

Sea salt

2 tablespoons strong brewed coffee or a shot of espresso

2 tablespoons balsamic vinegar

2 tablespoons whole coffee beans

2 teaspoons whole black peppercorns

1 tablespoon extra-virgin olive oil

1 (1¼-pound) flank steak

Combine the garlic, 1/4 teaspoon salt, brewed coffee, and vinegar in a small bowl; whisk together and reserve.

Place the coffee beans and peppercorns in a coffee grinder and pulse very briefly, until coarsely ground. (Alternatively, place them in a plastic bag and whack at them with a kitchen mallet or meat tenderizer mallet until coarsely cracked. This latter method is useful after a hard day at the office or when you would rather yell at your kids.)

Heat a grill pan over high heat or set gas grill on high.

When grill pan or gas grill are hot, rub oil on both sides of the steak, then press the coffee/peppercorn mixture evenly on both sides. Salt the steak to taste. Grill about 4 minutes per side to desired doneness—it is best medium-rare. Place the steak on a cutting board and let stand for 5 minutes.

Cut the meat very thinly across the grain on the diagonal and arrange over salad or greens of your choice. Heat the reserved dressing in a small pan and drizzle dressing over steak.

With Apologies to Joe's Special

PHASES 1–2, SERVES 3 TO 4

My favorite recipe to trick unsuspecting guests into consuming an impressive dose of greens, this dish can be refrigerated or frozen and reheated for a perfect one-pan dinner. It also makes a hearty weekend breakfast or brunch.

1 (10-ounce) package frozen chopped spinach or
1 (5-ounce) package fresh spinach, Swiss chard,
mustard greens, or kale

2 tablespoons extra-virgin olive oil

½ pound very lean grass-fed ground beef (preferred)
or ground turkey

2 cups sliced mushrooms (button, brown, cremini,
shiitake, or portobello)

1 cup chopped onion

1 garlic clove, minced, or ½ teaspoon garlic powder

1 tablespoon chopped fresh oregano,
or 1 teaspoon dried

1 teaspoon ground nutmeg

1 teaspoon ground coriander, or 2 tablespoons
chopped fresh coriander (cilantro)

½ teaspoon sea salt

2 dashes Tabasco or other hot sauce, or to taste

2 tablespoons Worcestershire sauce

4 omega-3 eggs

½ cup fat-free or low-fat cottage cheese

If using frozen spinach, thaw and drain in colander. If using bagged fresh greens, wash and trim if necessary.

In a large skillet or wok with a handle, heat the oil and then fry the meat with the mushrooms, onion, garlic, oregano, nutmeg, coriander, and salt over high heat until just done. If using beef, drain off any extra fat.

Add the spinach or other greens to the meat mixture and season with Tabasco and Worcestershire sauce. Cook for 2 to 3 minutes more, until greens are wilted but still bright green.

Beat the eggs, then pour over the spinach-meat mixture and stir until set lightly. Stir in the cottage cheese, adjust the seasonings, and serve.

PHASES 2–3

Reduce the amount of beef or turkey by one half and reduce eggs to 3.

PHASE 3

- Substitute two broken-up Boca Burger patties or other veggie burgers for the ground beef or turkey.

- Eliminate the cottage cheese, or replace it with 1/2 cup cubed avocado.

Peppered Garlic Pork Tenderloin

PHASES 1–3, SERVES 2 (AND LATER 4)

Vietnamese cooking is known for its contrast of salty and peppery tastes. This dish is a quick and easy way to prepare tenderloin, the only cut of pork you should consider using, thanks to its very low fat content. If possible, try to find pork from Neiman's Ranch or another sustainable pork farm.

2 garlic cloves, minced

1 teaspoon rubbed or powdered sage, or
    5 fresh sage leaves, finely chopped

1 teaspoon coarse sea salt

2 tablespoons black peppercorns, crushed

1 pork tenderloin (about ¾ pound)

1 tablespoon extra-virgin olive oil

Preheat the oven to 425°F.

Combine the garlic, sage, salt, and pepper to form a paste and press it into the pork tenderloin on all sides.

Put the oil in an ovenproof skillet (preferably cast iron) and heat over moderately high heat. When hot, put the pork in the pan and brown on all sides, using tongs to avoid piercing the meat. Remove from the heat and transfer pan to the oven. Roast for 20 minutes.

Transfer roast to a cutting board and let stand 5 to 10 minutes, then slice thinly on the diagonal. Arrange over salad greens, shredded cabbage, or half a bag of coleslaw mix.

Mustard Pork Tenderloin

PHASES 1–3, SERVES 2 (AND LATER 4)

Mustard and pork were made for each other, but there's a better way to enjoy the combo than on a hot dog. Try this method, and you'll never miss your "dog." Be sure to use whole-grain mustard, and the grainier the better.

1 garlic clove, crushed

1 tablespoon fresh rosemary, or 1 teaspoon dried

1 tablespoon fresh oregano, or 1 teaspoon dried

¾ cup whole-grain mustard

¼ cup or more red wine

¼ teaspoon each sea salt and crushed cracked black peppercorns

1 pork tenderloin (about ¾ pound)

Place all the ingredients except the pork in a large, sturdy resealable plastic bag and mix thoroughly. Add a little more wine if the marinade is not liquid enough. Add pork and reseal. Marinate in refrigerator at least 2 hours and preferably overnight.

Preheat a gas grill on high or use a grill pan.

When grill is hot, remove meat from the marinade and grill, turning frequently, until an instant-read meat thermometer indicates 170°F.

Remove to a carving board and let stand for 5 to 10 minutes, then slice thinly across the grain on the diagonal.

Serve over greens of your choice.

POULTRY

Do me a favor and go out of your way to find free-range chicken, which is much more flavorful and tender than conventionally raised birds. If you can't resist buying those huge packs of frozen breasts on special, I'll show you a few tricks to improve the tenderness of any chicken, in addition to slicing it very thin and stir-frying it.

Stir-Fried Chicken with Green Beans and Basil

PHASES 1–3, SERVES 2

Here's my secret way to get you to eat your veggies and want more. You can find the hot peppers in small cellophane bags, often in the Mexican spices section of your supermarket. Note that the cabbage is uncooked and chilled, which gives it a crunchy contrast to the rest of the dish.

2 tablespoons canola oil

1 teaspoon grated lime or lemon zest

1 tablespoon shredded or finely chopped fresh ginger

1 or 2 Thai red chiles or Japone or Chinese chiles, seeds removed, finely chopped; or ¼ teaspoon red pepper flakes

½ pound thinly sliced boneless chicken breasts or thighs

½ pound string beans, cut into 2- to 3-inch segments

10 or more whole basil leaves

½ cup light coconut milk

¼ to ½ teaspoon sea salt, or to taste

3 cups shredded cabbage, chilled

Heat the oil in a wok or deep skillet over high heat. Add the zest, ginger, and chiles, and stir briefly. Immediately add the chicken, the green beans, and basil. Stir-fry for 3 or 4 minutes, until done. Add the coconut milk and salt; mixture will be thick.

Serve over shredded cabbage.

VARIATIONS

- *Stir-Fried Beef with Green Beans and Basil:* Replace chicken with thinly sliced flank or round steak.

- *Stir-Fried Pork with Green Beans and Basil:* Replace chicken with thinly sliced pork tenderloin.

- *Stir-Fried Shrimp with Green Beans and Basil:* Replace chicken with shelled medium shrimp. You'll need about a pound of unpeeled shrimp to yield a half-pound peeled.

- After chicken is stir-fried, add 1/4 cup chunky all-natural peanut butter and 1/2 teaspoon curry powder mixed with 1 cup water; bring mixture to a boil and serve.

- *Stir-Fried Tempeh with Green Beans and Basil:* Substitute tempeh for the chicken.

- Instead of shredded cabbage, serve over your favorite greens.

- Instead of serving over shredded cabbage, prepare tofu shirataki noodles the Dr. G. way (see page 216). Divide onto plates and top with stir-fry.

Angelic Jungle Princess with Chicken

PHASE 1, SERVES 2

Here's my favorite Thai chicken recipe, which I adapted from the "Evil Jungle Prince with Chicken" served at Keo's on Oahu. I've taken the liberty of changing a lot of ingredients so that it can easily be made at home. Use as many chiles as you like, depending on your tolerance for heat.

2 to 6 small dried chiles,
  or ½ teaspoon red pepper flakes

1 teaspoon grated lemon or lime zest

2 tablespoons canola oil

½ cup light coconut milk

½ pound boneless chicken breasts or thighs,
  thinly cut into 2-inch strips

1 to 3 tablespoons fish sauce, or light soy
  sauce to taste

1 tablespoon lime or lemon juice

½ teaspoon sea salt

15 whole basil leaves

2 (8-ounce) packages tofu shirataki noodles,
  prepared the Dr. G. way (see page 216)

Grind the chiles and zest in a coffee or spice grinder. (Be sure to clean thoroughly after or use a grinder used only for savory foods.)

Heat the oil in a wok or deep skillet over medium-high heat, then sauté the chile mixture for 2 minutes, until very fragrant. Add the coconut milk and cook 2 minutes. Add the chicken and cook 3 minutes or until chicken is no longer pink. Reduce the heat to medium-low; add the fish sauce, lime or lemon juice, salt, and basil, stirring until basil is wilted.

Serve immediately over shirataki noodles.

PHASE 2

Replace the noodles with lightly sautéed or steamed thinly sliced cabbage.

PHASE 3

Replace the noodles with 2 cups raw thinly sliced cabbage or cooked or raw fresh asparagus spears, green beans, and mushrooms.

Fried Chicken with Almond "Breading"

PHASES 1–3, SERVES 2 (AND LATER 4)

Believe it or not, it's not the oil used for frying that's killing you; instead, it's old oil that's turned into trans fats, plus the breading made from flour. Here's my version of the juiciest, most flavorful fried chicken you've ever tasted.

1 cup buttermilk

½ teaspoon Tabasco or other hot sauce

2 skinless, boneless free-range chicken breasts

½ cup almond meal

¼ teaspoon sea salt

¼ teaspoon cracked black peppercorns

4 tablespoons extra-virgin olive oil

Fresh lemon (optional)

Combine the buttermilk and Tabasco in a sturdy resealable plastic bag or flat casserole dish. Pound the chicken breasts under a piece of plastic wrap with a wooden mallet or back of a heavy spoon until about 1/2 inch thick. Place the breasts in the buttermilk mixture and soak at room temperature for 20 to 30 minutes.

Meanwhile, mix the almond meal, salt, and pepper and place in another resealable plastic bag.

Drain the chicken breasts and immediately drop one into the bag with the almond meal mix and shake to cover. Remove and repeat with second breast.

Heat the oil in a heavy skillet over medium-high heat. Add the chicken and cook about 4 minutes on first side. Lower the heat to medium, turn the chicken, and cook 4 minutes on other side, or until done.

Remove the chicken to a carving board to sit for 2 minutes. Slice into 1/2 inch strips and serve over lettuce or other greens. Drizzle with oil and a squeeze of lemon juice.

Pesto Chicken: Prepare the following sauce and pour over the fried chicken: In blender or bowl of a small food processor, combine 1 cup basil (or Italian or curly parsley or cilantro), 1 garlic clove, and 4 tablespoons pine nuts or walnuts. Pulse, adding 1/2 cup extra-virgin olive oil and processing until finely minced. Add extra oil if necessary to thin the sauce and, with motor running, add a 1-inch cube of Parmigiano-Reggiano cheese, if desired.

Dr. G's South by Southwest Chicken

PHASES 1–3, SERVES 4 (AND LATER 8)

Having lived in Baltimore, Washington, D.C., Atlanta, and now Palm Springs, I was delighted to win second place in the Old El Paso cooking contest with this recipe, that pays healthy homage to both regions.

1 red onion, chopped

1 fresh pasilla or Anaheim chile, seeded and
    chopped, or 1 (4-ounce) can green chiles

1 tablespoon pure chile powder

1 tablespoon Old Bay Seasoning

Juice of 4 lemons

2 garlic cloves, minced

1 teaspoon sea salt

1 teaspoon ground ginger

2 teaspoons curry powder

½ cup extra-virgin olive oil

1 free-range chicken, cut into pieces,
    or 2 thighs and 2 breasts, boned

Put all ingredients except chicken in a large, sturdy resealable plastic bag or shallow ovenproof casserole and mix well. Add the chicken and marinate for 4 hours or overnight in the refrigerator.

Preheat the oven to 350°F.

Put the chicken and marinade in a shallow ovenproof baking dish and bake, covered with foil, for 45 minutes, basting occasionally. Uncover and bake 10 minutes more. (Alternatively, remove from pan after 45 minutes and finish in grill pan or on grill for 10 minutes to crisp.) Serve with salad or thinly slice and place on prepared shirataki noodles (see page 216).

Chicken Under a Brick

PHASES 1–3, SERVES 2 (AND LATER 4)

Sometimes the simplest is the best. This is my homage to Marioni's, a tiny, locals-only trattoria on a back street in Florence, where construction workers mingle with office workers decked out in Armani suits, all ordering this dish!

1 lemon, thinly sliced

6 garlic cloves, unpeeled

1 whole chicken breast, breastbone removed, chicken flattened

Extra-virgin olive oil

¼ teaspoon sea salt

¼ teaspoon or less red pepper flakes

8 sprigs fresh thyme

10 sprigs fresh oregano

¼ teaspoon cracked black peppercorns

A brick wrapped in foil

Juice of 1 lemon

Shaved Parmigiano-Reggiano cheese, for garnish

Preheat the oven to 400°F.

Arrange the lemon and garlic on the bottom of a shallow roasting pan. Place the chicken, rib side down, in the pan; brush with oil, sprinkle with salt and red pepper flakes, and cover with thyme, oregano, and pepper. Place the brick on top of the chicken and roast for 20 minutes. Remove the brick and roast uncovered for an additional 25 minutes, or until nicely browned.

Remove to a carving board, allow to rest for 5 to 10 minutes, then slice into 1/2-inch-thick slices. Arrange the chicken slices on a bed of greens. Drizzle on olive oil, and squeeze lemon juice over everything. If desired, sprinkle with shaved cheese.

This same recipe can be made on the stovetop, with equally superb results. Omit the sliced lemon; put 2 tablespoons olive oil in a deep skillet and heat over medium-high heat. Add the chicken and herbs and press down with the brick. After 10 minutes, remove the brick, flip the chicken, and recover with the brick for an additional 10 minutes, or until done.

Chicken and Black Soybean Quesadilla Casserole

PHASE 1, SERVES 4

Most casseroles are based on pasta or other starches, but not this one, which I surgically remodeled from a high-starch version in *Cooking Light.* You can make it ahead and refrigerate or freeze until needed. Be sure to use black soybeans, not black beans. Queso blanco is a Mexican cheese that doesn't melt. Or substitute shredded soy Jack cheese.

1 cup thinly sliced red onion

5 garlic cloves, minced

1 cup chopped Soyrizo or chorizo sausage

1 pasilla or Anaheim chile, roasted, seeded, and
    chopped, or 1 (4-ounce) can chopped green chiles

2 cups shredded or cubed cooked chicken

Sea salt and cracked black peppercorns to taste

1 (15-ounce) can black soybeans, drained

1 cup chicken broth

1½ cups fresh or canned salsa, plus extra for serving

Canola cooking spray or 1 tablespoon
    extra-virgin olive oil

6 low-carb, high-fiber tortillas, cut into 1-inch strips

1 cup shredded queso blanco cheese

Preheat the oven to 450°F.

In a nonstick skillet or wok over medium-high heat, sauté the onion and garlic for 5 minutes. Add the Soyrizo or chorizo and chile and cook for several minutes, until browned. Add the chicken and stir for a few minutes. Season with salt and pepper, then transfer to a bowl and stir in the beans.

In same pan, bring the chicken broth and salsa to a boil. Reduce the heat and simmer for 5 minutes, until the sauce is reduced, stirring occasionally.

Spray an 11-by-7-inch casserole dish with cooking spray or lightly coat with olive oil. Arrange a layer of tortilla strips on the bottom and spoon half of the chicken-chile mixture over them. Top with the remaining tortilla strips and follow with a second layer of chicken mixture. Pour the salsa mixture over the casserole and sprinkle the cheese on top. Bake for 15 minutes, until the cheese browns slightly. Serve with salad of your choice, or simply shred romaine lettuce and top with a portion of casserole, and extra salsa on the side. (If wrapping and freezing for later use, thaw and bake 30 minutes before serving.)

PHASE 2

· Reduce tortillas to 3.

· Replace the chicken with Lightline ground beef-style vegemeat or tempeh, cubed.

PHASE 3

Eliminate the chicken and add 1 cup chopped spinach or other greens. This version serves three instead of four.

FISH

Other than overdone chicken breasts, nothing is worse than dried-out, over-cooked fish. In my opinion, the best place for a good fish is atop a cold or warm salad. These recipes offer a surefire way of getting fish into your diet without the fuss. Try to find line-caught wild fish, if possible.

Simply Grilled or Pan-Fried Fish

PHASES 1–3, SERVES 2

Even Costco has wild Copper River salmon fillets in the freezer section. My favorite Cajun seasoning is Emeril's.

3 tablespoons extra-virgin olive oil

1 teaspoon Old Bay or Cajun seasoning

2 (4- to 6-ounce) fillets of salmon, tuna, grouper, or other "meaty" fish

Canola or olive oil nonstick cooking spray or 1 tablespoon extra-virgin olive oil

Combine the oil and seasoning in a large, sturdy resealable plastic bag and mix well. Add the fish and marinate at room temperature for 20 minutes.

Spray a grill pan with cooking spray or coat with oil and place over high heat.

If using tuna or salmon, sear for 1 minute on each side for very rare. Or, until you get used to seared rare salmon, cook for 2 to 4 minutes on each side. Slice into 1/4-inch strips and serve over salad of your choice.

Commander Gundry's Pecan Fish

PHASE 1, SERVES 4

Several years before his untimely death, I questioned chef Jamie Shannon while dining at his incomparable New Orleans restaurant, Commander's Palace, about a dish that combined fish and nuts. That discussion sparked my take on a great fish dish. In this recipe, nothing can replace the matchless taste of butter. If using trout, don't remove the skin.

2 cups fresh pecans or pecan pieces

1 cup almond meal or soy flour

2 tablespoons Cajun seasoning, such as Emeril's, plus additional for dusting fish

1 omega-3 egg

1 cup unsweetened soymilk

4 catfish, tilapia, or rainbow trout fillets

2 tablespoons grass-fed butter (see Note)

¼ cup extra-virgin olive oil

2 teaspoons fresh lemon juice

1 tablespoon Worcestershire sauce

2 tablespoons chopped fresh thyme, or 1 tablespoon dried

Sea salt and cracked black peppercorns to taste

Combine 1 cup of the pecans, the almond meal or soy flour, and the Cajun seasoning in a food processor. Pulse until finely ground but not smooth, then put into a shallow, wide plate.

Whisk the egg and soymilk in a large bowl.

Lightly dust each fish piece with additional Cajun seasoning, then dip into the egg wash. Place the fish on the pecan mixture and coat both sides, shaking off excess.

Heat the butter in a large nonstick skillet over high heat until bubbling, then add the fish and cook for 30 seconds. Reduce the heat to medium and cook for 2 minutes more. Turn the fillets and cook for 2 to 3 more minutes, until evenly browned. Remove to serving plates.

In the same skillet, place the olive oil and remaining 1 cup pecans; heat over high heat for 2 minutes, until bubbly, stirring constantly. Add the lemon juice, Worcestershire sauce, thyme, and salt and pepper to taste, and cook another minute. Top each fish fillet with a little pecan mixture.

PHASE 2

Place a generous amount of sautéed greens or fresh lettuce on a plate, top with a fillet, and then spoon pecans over top. Add additional olive oil and lemon juice, if desired.

PHASE 3

Commander Gundry's Pecan Tempeh: Substitute slices of tempeh for the fish and arrange over fresh greens or lettuces, as above.

NOTE: Although butter will make only a rare appearance in Diet Evolution, in some recipes its special flavor is occasionally the only ingredient that will do. However, most American butter is made from the milk of cows that consume grain and soybeans, and is therefore high in inflammation-causing omega-6 oils. For a delicious alternative full of omega-3 fatty acids, use only butter from grass-grazed cows. In addition to Irish and French butter, some American brands, such as PastureLand and Straus Family Creamery, are good choices.

COOKED VEGETABLES

I've tried to include as many raw or lightly stir-fried vegetables and greens as possible in other dishes, but there are also a few additional cooked vegetable dishes you must try.

Sag Paneer

PHASES 1–3, SERVES 2

I love Indian cooking, but so many of the dishes are unhealthy messes. Here's one of my favorites—without the high fat content.

>   1 teaspoon curry powder, or to taste
>
>   2 tablespoons extra-virgin olive oil
>
>   1 (10-ounce) package frozen chopped spinach, thawed and drained
>
>   1 teaspoon sea salt
>
>   1 cup low-fat cottage cheese or low-fat ricotta

In a skillet or wok, combine the curry powder and oil and cook over low heat for 3 minutes, until fragrant.

Add the spinach and salt. Raise the heat to medium and cook, covered, for 5 minutes. Stir in the cheese and heat thoroughly. Serve as a side dish or as a base for one of the sliced meat recipes.

Brussels Sprouts
You'll Eat

PHASES 1–3, SERVES 2

The cabbage family has some of the most important anticancer compounds known to humans, yet most of us eschew all but broccoli. Funny little mini-cabbages, Brussels sprouts are usually ignored because they smell bad when they're overcooked, but here's a way to fix them that guarantees you'll enjoy 'em!

1 pound Brussels sprouts

3 tablespoons extra-virgin olive oil

½ cup raw hazelnuts or walnuts

¼ teaspoon sea salt

¼ teaspoon cracked black peppercorns

1 teaspoon roasted sesame oil (optional)

Trim the tough stem ends of the Brussels sprouts and remove any discolored leaves. Using a food processor fitted with the slicing blade, place the sprouts in the tube and slice thin. (Alternatively, slice as thin as possible by hand or use a mandoline.)

Heat 1 tablespoon of the oil in a skillet or wok over medium-high heat. Add the nuts and cook for 5 minutes, until fragrant. Remove and reserve. Add 2 remaining tablespoons oil and the sprouts to the skillet; sauté, stirring for 7 or 8 minutes, until browned.

Return the nuts to the pan and mix with the Brussels sprouts. Season with salt and pepper. Pour the sesame oil over just before serving. Serve alone or use as a base for meat or poultry dishes.

VARIATION

Nutty Green Beans: Substitute 1 pound green beans for the Brussels sprouts. Trim, but don't slice in food processor. After mixing with the nuts, sprinkle juice of half a lemon over the top and serve.

PHASE 3

Combine the ingredients and eat raw! Don't wrinkle up your nose until you've tried it this way.

Balsamic Asparagus

PHASES 1–3, SERVES 2

I can never get enough asparagus! As the growing tips of a tall fernlike plant, asparagus is the ultimate in health food. But let's mix it up a bit, shall we? Rather than cutting off those tough asparagus ends, bend them near the ends until the woody parts snap off.

3 tablespoons extra-virgin olive oil

1 tablespoon balsamic vinegar

1 (1-pound) bunch asparagus, snapped at bending point, rinsed, and dried

¼ teaspooon sea salt, or to taste

Cracked black peppercorns to taste

1–2 tablespoons seeds or nuts (optional; see Note)

Shaved Parmigiano-Reggiano cheese (optional)

Heat the oil in large nonstick skillet over medium-high heat; add the vinegar and stir. Add the asparagus and stir-fry until crisp-tender, 5 to 7 minutes. Add salt and pepper to taste, then top with nuts and cheese, if desired.

NOTE : If desired, add to the hot pan a handful of any nuts or seeds and 1 teaspoon additional olive oil. Stir-fry for 2 minutes, or until fragrant. My favorites here are walnuts, pine nuts, or pumpkin seeds, but mix it up. You should also try sesame seeds, pecans, or pistachios.

PHASE 3

Finely chop the asparagus, combine all the ingredients, and serve raw. Don't laugh—this may soon become your favorite way to eat asparagus!

Grilled Green Beans

PHASES 1–2, SERVES 2

If you've ever grown tired of green beans, you probably didn't know that grilling brings out the best in them. If possible, use the small, young beans that the French call *haricot verts.* This cooking method makes virtually any vegetable taste better.

1 pound green beans, stem-ends trimmed

3 tablespoons extra-virgin olive oil

1 tablespoon Cajun seasoning, or
    2 tablepoons chopped fresh rosemary,
    sage, thyme, or oregano

Sea salt to taste

Cracked black peppercorns to taste

Combine the beans, oil, and seasoning in a resealable plastic bag or shallow baking dish and seal or cover; marinate for 30 minutes or refrigerated overnight.

Heat a grill pan over high heat. Distribute the green beans over the grill pan, stirring or shaking the pan constantly, until the beans start to brown, 3 or 4 minutes. Season with salt and pepper, and serve immediately.

PHASE 3

Coarsely chop the raw green beans, mix with the herbs, and sprinkle with balsamic vinegar or lemon juice. Serve raw.

Roasted Cauliflower with Sage

PHASES 1–2, SERVES 4

One of my all-time favorite flavor combinations is sage and brown butter, which usually accompanies pasta in Italy. Instead, enjoy "pasta" with minimal carbs by using cauliflower in a new way: roast it!

1 head cauliflower, or 1 (16-ounce) bag florets

2 tablespoons grass-fed butter (see Note, page 255)

¼ cup extra-virgin olive oil

3 tablespoons chopped fresh sage, or
    1 tablespoon dried

¼–½ teaspoon sea salt, or to taste

¼–½ teaspoon cracked black peppercorns, or to taste

Preheat the oven to 400°F.

If using whole cauliflower, cut into florets and cut stems into 1-inch pieces.

Brown the butter in a large ovenproof skillet over medium-high heat for 2 minutes, until bubbly. Add the olive oil and sage; stir to combine. Add the cauliflower, salt, and pepper; stir to coat cauliflower with seasonings. Immediately transfer to the oven and bake for 30 minutes, turning and stirring pieces at least once; the cauliflower should have some brown edges. Serve.

PHASE 3

Omit the butter. Place the cauliflower in a food processor and pulse until pea-size. Place in a bowl with the chopped sage, olive oil, salt, and pepper. Mix well and serve raw, with shaved Parmigiano-Reggiano cheese on top.

DESSERTS

I'd prefer you finish a meal with some berries, a handful of walnuts, or a piece of dark chocolate with more than 70 percent cocoa, but I also understand that you occasionally crave something more. The recipes that follow will help you, as they did me, through Phase 1 and into Phase 2, but please have them only as a special treat in Phase 3.

Dr. G.'s Berry Ice Dream

PHASES 1 *(AFTER THE FIRST TWO WEEKS)* **AND 2,** SERVES 1

For the first year of my own Diet Evolution, this treat got me through many a night! I recommend BioChem protein powder.

½ to 1 scoop low-carb vanilla protein powder

½ cup or more unsweetened vanilla soymilk

1 cup frozen blueberries or frozen mixed berries

Pinch of stevia or other no-calorie sweetener (optional)

In a cereal bowl, dissolve the protein powder in the soymilk. Stir in the frozen berries and continue to stir until they freeze the soymilk mixture into "soft ice cream." Add a little more soymilk if the mixture is too thick. This is sweet enough for me as is, but if you want it sweeter, sprinkle a bit of stevia to taste. Just remember to pinch the packet (see "Pinch the Packet" on page 87).

Protein-Powered Mint Chocolate Chip Ice Dream

PHASES 1 *(AFTER THE FIRST TWO WEEKS)* **AND 2,** SERVES 1 OR 2

Here's my version of milk-free soft-serve ice cream. It may take you a few tries to get a creamy consistency; just keep tasting and blending. Start with two packets of stevia and add more only if you need it.

1 unripe banana, peeled and frozen (see Note)

1 tablespoon unsalted tahini (preferably raw)

1 tablespoon non-Dutch-processed cocoa powder

1 scoop chocolate or vanilla whey protein powder

3 fresh mint leaves, or 4 drops mint extract

2 packets stevia or other no-calorie sweetener

3 cups (approximately) ice cubes

Unsweetened soymilk to thin (optional)

1 tablespoon raw cocoa nibs

Place the banana, tahini, cocoa powder, protein powder, mint leaves or extract, sweetener, and 1 cup of ice cubes into a blender. Use ice-crushing mode or slow speed, pulse until the ice begins to fragment, then gradually increase the speed, adding more ice cubes to make a very thick consistency. Add a bit of soymilk, a tablespoon at a time, if it's too thick to mix well. Blend at high speed until the mixture reaches a creamy consistency. Add the cocoa nibs and blend briefly to distribute but do not grind. Serve immediately in bowls.

NOTE: Buy bananas green, peel them, and freeze them in resealable plastic bags so you can use them one at a time.

Almond Chocolate Chip Surprise Cookies

PHASES 1–3, MAKES 12 TO 14 COOKIES

When you must have a cookie, try these large, healthy, tasty treats—full of protein, nuts, dark chocolate, and hidden veggies. Have just one or two; they're not low in calories.

½ cup almond butter

2 omega-3 eggs

¼ cup extra-virgin olive oil

⅓ cup low-fat ricotta or cottage cheese

1 teaspoon baking soda

½ teaspoon sea salt

⅓ cup Splenda for baking

½ teaspoon vanilla extract

1 teaspoon ground cardamom

½ cup vanilla whey powder

½ cup shredded, unsweetened coconut

1 cup chopped spinach

1 cup grated zucchini

1½ cups almond flour

½–1 cup chopped (70 percent or more cocoa) dark chocolate

Canola oil cooking spray

Preheat the oven to 375°F.

In an electric mixer, blend the almond butter, eggs, oil, ricotta, baking soda, salt, Splenda, vanilla, and cardamom. Slowly add the whey powder, coconut, spinach, zucchini, and almond flour, until mixed. On low speed, or by hand, blend in the chocolate pieces.

Grease a nonstick cookie sheet with cooking spray and drop generous spoonfuls of dough onto the sheet. Cookies will look lumpy. Bake on the middle rack of the oven for 6 to 8 minutes. They will still look a little under-cooked. Remove with a spatula to cool briefly on a rack.

Eat cookies warm from the oven. They will remain fresh for a few days, or refrigerate or freeze leftovers.

NOTES

Chapter 1: YOUR GENES ARE RUNNING THE SHOW

1. Stern, K., and McClintock, M. 1998. "Regulation of ovulation by human pheromones." *Nature* 392:177–179.

2. Myslobodsky, M. 2003. "Gourmand savants and environmental determinants of obesity." *Obesity Reviews* 4(2):121–28.

3. Douglas, J. L., and Goldhamer, A. 2003. *The Pleasure Trap.* Summertown, TN: Healthy Living Publishing.

4. Delgado, J. M. R. 1969. *Physical Control of the Mind: Towards a Psychocivilized Society.* New York: Harper and Row.

5. Morgan, J. P., and Zimmer, L. 1997. "The social pharmacology of smokeable cocaine." In *Crack in America: Demon Drugs and Social Justice*, C. Reinarman, and H. G. Levine, eds, 131–170. Sacramento: Regents of University of California Press.

6. Phillips, H. 2003. "The Pleasure Seekers." *The New Scientist* 11 (October): 38–43.

7. *Nutritional Requirements of Non-Human Primates*, 2nd ed. 2003. National Research Council, Washington, DC: National Academic Press.

8. Gibbon, J., and Cherfas, J. 2001. *The First Chimpanzee.* New York: Barnes and Noble Books.

9. Cai, Y., and Feng, W. 2006. "Famine, social disruption, and miscarriage: Evidence from Chinese survey data." *CSDE Working Paper 04-06*, Seattle: University of Washington Press.

10. Catalano, R., and Bruckner, T. 2006. "Secondary sex ratios and male lifespan: Damaged or culled cohorts?" *Proceedings of the National Academy of Science* 10 (January), 1073.

11. Louks, A. B., Verdun, M., and Heath, E. M. 1998. "Low energy availability, not stress of exercise, alters LH pulsatility in exercising women." *Journal of Applied Physiology* 84:3246.

12. Gallou-Kabani, C., and Junien, C. 2005. "Nutritional epigenomics of metabolic syndrome." *Diabetes* 54:1899–1906.

13. Masters, J. R. W., and Palsson, B. (eds.). 1998. "Human Cell Culture," Vol. 1, *Cancer Cell Lines Part 1.* Springer, NY: Kluwer Academic.

14. Blanc, S., Normand, S., et al. 2000. "Fuel homeostasis during physical inactivity induced by bed rest." *Journal of Clinical Endocrinology and Metabolism* 85:2223–33.

15. Knez, W. L., Coombes, J. S., and Jenkins, D. G. 2006. "Ultra-endurance exercise and oxidative damage: Implication for cardiovascular health." *Sports Medicine* 36(5):429–41.

16. Cleave, T. L. 1974. *The Saccharine Disease.* Bristol, UK: John Wright and Sons.

17. Price, W. 1997. *Nutrition and Physical Degeneration.* New York: Keats Publishing.

18. Lindsay, R. S., Cook, V., et al. 2002. "Early excess weight gain by children in the Pima Indian population." *Pediatrics* 109:33–39m.

19. Knowler, W. C., Saad, M. F., et al. 1991. "Obesity in the Pima Indians: Its magnitude and relationship with diabetes." *American Journal of Clinical Nutrition* 53:1543–51.

20. Wolk, A., Gridley, G., et al. 2001. "A prospective study of obesity and cancer risk." *Cancer Causes and Control* 12:13–21.

21. Calle, E. E., Rodriguez, C., et al. 2005. "Overweight, obesity and morbidity from cancer in a prospectively studied cohort of U.S. adults." *New England Journal of Medicine* 348:1625–38.

22. Levine, B. D., and Thompson, P. D. 1997. "Marathon maladies." *New England Journal of Medicine* 352:1516–18.

23. Bruungaard, H., Hartkopp, A., et al. 1997. "In vivo cell-mediated immunity and vaccination response following prolonged intense exercise." *Medical Science in Sports and Exercise* 29(9):1176–81.

24. Haley, R. W. 2003. "Excess incidence of ALS in young gulf war veterans." *Neurology* 61:750–56.

25. Srivastava, A., and Kreiger, N. 2000. "Relation of physical activity to risk of testicular cancer." *American Journal of Epidemiology* 151:78–87.

26. Hoenig, M. J., Botma, A., et al. 2007. "Long-term risk of cardiovascular disease in 10-year survivors of breast cancer." *Journal National Cancer Institute* 99:363-75.

27. Fosså, S. D., Gilbert, E., et al. 2007. "Noncancer causes of death in survivors of testicular cancer." *Journal National Cancer Institute* 99:533-44.

28. Korol, D. L., and Gold, P. E. 1998. "Glucose, memory, and aging." *American Journal of Clinical Nutrition* 67:764s-71s.

29. Saposky, R. M. 2001. *A Primate's Memoir.* New York: Simon & Schuster.

30. Thorpe, K. E., and Howard, D. H. 2006. "The rise in spending among Medicare beneficiaries: The role of chronic disease prevalence and changes in treatment intensity." *Health Affairs* 25:w378-w88.

31. Flegal, K. M., Graubard, B. I., et al. 2005. "Excess deaths associated with underweight, overweight, and obesity." *Journal of American Medical Association,* 293:1861-67.

32. Manson, J. E., Colditz, G. A., et al. 1990. "A prospective study of obesity and risk of coronary heart disease in women." *New England Journal of Medicine* 322:882-89.

33. Kuk, J. L., Katzmarzyk, P. T., et al. 2006. "Visceral fat is an independent predictor of all-cause mortality in men." *Obesity* 14:336-41.

34. Ricci, E., Smallwood, S., et al. 2006. "Electrophysiological characterization of left ventricular myocytes from obese Spraque-Dawley rats." *Obesity* 14:778-86.

Chapter 2: WE ARE WHAT WE EAT

1. Knott, C. D. 1998. "Changes in orangutan caloric intake, energy balance, and ketones in response to fluctuating fruit availability." *International Journal of Primatology* 19:1061-79.

2. Brockman, D. K., van Schaik, C. P. (eds.). 2005. *Seasonality in Primates.* New York: Cambridge University Press.

3. Popovich, D. G., and Dierenfeld, D. 1997. "Gorilla nutrition." In *Management of Gorillas in Captivity: Husbandry Manual, Gorilla Species Survival Plan,* J. Ogden, and D. Wharon, eds. Silver Springs, Md.: American Association of Zoos and Aquariums.

4. Jenkins, D. J. A., and Kendall, C. W. C. 2006. "The Garden of Eden: Plant-based diets, the genetic drive to store fat and conserve cholesterol, and implication for epidemiology in the 21st century." *Epidemiology* 17(2):128-30.

5. Blumineshine, R. J., and Cavallo, J. A. 1992. "Scavenging and human evolution." *Scientific American* 267:90-96.

6. Gundry, S. R. 1972. *Human Biological and Social Evolution.* Divisional IV Honors Thesis, Yale University.

7. Leutenegger, W., and Lubach, G. 2005. "Sexual dimorphism, mating system, and effect of phylogeny in De Brazza's monkey." *American Journal of Primatology* 13:171-79.

8. Liang, Y., Steinbach, G., et al. 1987. "The effect of artificial sweeteners on insulin secretion." *Hormonal and Metabolic Research* 19(6):233-38.

9. Dhingra, R., Sullivan., L. M., et al. 2007. "Soft drink consumption and risk of developing cardiometabolic risk factors and the metabolic syndrome in middle-aged adults in the community." *Circulation 2007* 116:480-88.

10. Davidson, T., and Swithers, S. E. 2004. "A Pavlovian approach to the problem of obesity." *International Journal of Obesity* 28:933-35.

11. Ephraim, R. 2001. *Wise Traditions in Food, Farming, and the Healing Arts.* Washington, D.C.: Weston A. Price Foundation.

12. Leeson, C. P. M., Kattenhorn, M., et al. 2001. "Duration of breast feeding and arterial distensibility in early adult life: population based study." *British Medical Journal* 322:643-47.

13. Bidlack, W. R., Omaye, S. T., et al. (eds.). 2000. *Phytochemicals as Bioactive Agents.* Palm Beach: CRC Press.

14. Meskin, M. S., Bidlack, W. R., et al. (eds.). 2003. *Phytochemicals: Mechanisms of Action.* Palm Beach: CRC Press.

15. Murata, M. 2000. "Secular trends in growth and changes in eating patterns of Japanese children." *American Journal of Clinical Nutrition* 72:1379S-83S.

16. Samaras, T. J., and Elrick, H. 2002. "Height, body size and longevity: Is smaller better for the human body?" *Western Medical Journal* 17:206-08.

17. Campbell, S. 1978. "Noah's ark in tomorrow's zoo: Animals are a comin', two by two." *Smithsonian* 8:42-50.

18. Sho, H. 2001. "History and characteristics of Okinawan longevity food." *Asia Pacific Journal of Clinical Nutrition* 10:159-64.

19. Ghirardini, M. P., Carli, M., et al. 2007. "The importance of a taste: a comparative study on wild food plant consumption in twenty-one local communities in Italy." *Journal of Ethnobiology Ethnomedicine* 3:22.

20. Logan, A. C. 2004. "Omega-3 fatty acids and major depression: A primer for the mental health professional." *Lipids in Health and Disease* 3:25.

21. Peet, M., and Horrobin, D. F. 2002. "A dose-ranging study of the effects of Ethyl-Eicosapentaenoate in patients with ongoing depression despite apparently adequate treatment with standard drugs." *Archives of General Psychiatry* 59:913-19.

22. Cochet, N., and Georges, B. 1999. "White adipose tissue fatty acids of alpine marmots during their yearly cycle." *Lipids* 34:275-81.

23. Wyshak, G., and Frisch, R. E. 1982. "Evidence for a secular trend in age of menarche." *New England Journal of Medicine* 306:1033-35.

24. Altmann, J., and Alberts, S. 1987. "Body mass and growth rates in a wild primate population." *Oecologia* 72:15-29.

25. *US Teenage Pregnancy Statistics,* National and State Trends. 2006. New York: Guttmacher Institute.

26. Saposky, R. M. 2001. *A Primate's Memoir.* New York: Simon & Schuster.

Chapter 3: CHANGING THE MESSAGE

1. Dansinger, M. L., Gleason, J. A., et al. 2005. "Comparison of the Atkins, Ornish, Weight Watchers, and Zone diets for weight loss and heart disease reduction." *Journal of the American Medical Association* 293(1):43-53.

2. Fleming, R. M. 2002. "The effect of high, moderate, and low-fat diets on weight loss and cardiovascular disease risk factors." *Preventive Cardiology* 5(3):110-18.

3. Freedman, M. R., King, J., and Kennedy, E. 2001. "Popular diets: A scientific review." *Obesity Research* 9:15-55.

4. Walford, R., Mock, D., et al. 1999. "Physiologic changes in humans subjected to severe, selective caloric restriction for 2 years in Biosphere 2: Health, aging and toxicological perspectives." *Toxicological Sciences* 52:61-65.

5. Hall, D., and Quaglia, R. 1998. *Great Aspirations.* Kansas City: Universal Press Syndicate.

Chapter 4: THE DIET AT A GLANCE

1. Walford, R. 2000. *Beyond the 120-Year Diet.* New York: Four Walls Eight Windows.

Chapter 5: THE FIRST TWO WEEKS

1. Stefansson, V. 1956. *The Fat of the Land.* New York: Macmillan.

2. Halton, T. L., and Hu, F. B. 2004. "The effects of high protein diets on thermogenesis, satiety, and weight loss: a critical review." *Journal of the American College of Nutrition* 23:373-85.

3. Maffiuletti, N.A., Agosti, F., et al. 2005. "Changes in body composition, physical performances and cardiovascular risk factors after a 3-week integrated body weight reduction program and after 1-year followup in severely obese men and women." *European Journal of Clinical Nutrition* 59:685-94.

4. Neese, R. M., and Williams, G. C. 1996. *Why We Get Sick.* New York: Vintage.

5. Gundry, S. R., and Ehrman, W. J. 2006. "Diet and supplement effects on lipid subfractions and inflammatory markers in the metabolic syndrome: Change is possible." *Atherosclerosis* 7 (Suppl. 3):440.

6. Khan, A., Safdar, M., et al. 2003. "Cinnamon improves glucose and lipids of people with type 2 diabetes." *Diabetes Care* 26:3215-18.

7. Mang, B., Wolters, M., et al. 2004. "Effects of a cinnamon extract on plasma glucose, HBA1C, and serum lipids in diabetes mellitus type 2." *European Journal of Clinical Investigation* 36(5):340-44.

8. Shindea, U. A., Sharma, G., et al. 2004. "Insulin sensitizing action of chromium picolinate in various experimental models of diabetes mellitus." *Journal of Trace Elements in Medicine and Biology* 18(1):23-32.

Chapter 6: WHAT'S OFF THE MENU?

1. Freedman, M. R., King, J., and Kennedy, E. 2001. "Popular diets: A scientific review." *Obesity Research* 9:15-55.

2. Walford, R., Mock, D., et al. 1999. "Physiologic changes in humans subjected to severe, selective caloric restriction for 2 years in Biosphere 2: Health, aging and toxicological perspectives." *Toxicological Sciences* 52:61-65.

3. Gundry, S. R., and Ehrman, W. J. 2005. "Effect of nutritional and supplement intervention on weight, HDL cholesterol, and metabolic syndrome in the

elderly with coronary artery disease." *Journal of Nutrition, Health and Aging* 9(3):153.

Chapter 7: THE TEARDOWN CONTINUES

1. Goyarza, P., Malin, D.H., et al. 2004. "Blueberry supplemented diet: Effects on object recognition, memory and nuclear factor kappa B levels in aged rats." *Nutritional Neuroscience* 7(2):75–83.

2. *"Grizzly and Black Bear Diet Throughout the Season.* © Ward Cameron 2005. Accessed 8/28/07; www.MountainNature.com/Wildlife/Bears/Bear Diet.htm.

3. Calle, E. E., Rodriguez, C., et al. 2005. "Overweight, obesity and morbidity from cancer in a prospectively studied cohort of U.S. adults." *New England Journal of Medicine* 348:1625–38.

4. Gundry, S. R., and Ehrman, W. J. 2006. "Diet and supplement effects on lipid subfractions and inflammatory markers in the metabolic syndrome: Change is possible." *Atherosclerosis* 7 (Suppl. 3):440.

5. Gundry, S. R., and Ehrman, W. J. 2005. "Raising HDL with diet and supplements: accomplishing the impossible." *Diabetes and Vasular Research Journal* 3(2):134.

6. Gundry, S. R., and Ehrman, W. J. 2006. "Control of Lp(a) with diet and supplements." *Atherosclerosis* 7 (Suppl. 3):530.

7. deLorgeril, M., Salen, P., et al. 1999. "Mediterranean diet, traditional risk factors, and the rate of cardiovascular complications after myocardial infarction: The final report of the Lyon Diet Heart Study." *Circulation* 99:779–85.

8. Kiecolt-Glaser, J. K., Belury, M. A., et al. 2007. "Depressive symptoms, Omega-6:Omega-3 fatty acids and inflammation in older adults." *Psychosomatic Medicine* 69(3):217–24.

9. Northcott, C. A., and Watts, S. W. 2003. "Low MG enhances arterial spontaneous tone via phosatidylinositol 3-kinase in DOCA-salt hypertension." *Hypertension* 43:125–29.

10. Van der Tempel, H., Tulleken, J. E., et al. 1990. "Effects of fish oil supplementation in rheumatoid arthritis." *Annals of the Rheumatic Diseases* 49(2):76–80.

11. Hill, A. M., and Buckley, J. D. 2007. "Combining fish-oil supplements with regular aerobic exercise improves body composition and cardiovascular disease risk factors." *American Journal of Clinical Nutrition* 85:1267–74.

1. Gundry, S. R., and Ehrman, W. J. 2006. "Diet and supplement effects on lipid subfractions and inflammatory markers in the metabolic syndrome: Change is possible." *Atherosclerosis* 7 (Suppl. 3):440.

2. Gundry, S. R., and Ehrman, W. J. 2005. "Effect of nutritional and supplement intervention on weight, HDL cholesterol, and metabolic syndrome in the elderly with coronary artery disease." *Journal of Nutrition, Health and Aging* 9(3):153.

3. Gundry, S. R., and Ehrman, W. J. 2005. "Raising HDL with diet and supplements: accomplishing the impossible!" *Diabetes and Vascular Research Journal* 3(2):134.

4. Gundry, S. R., and Ehrman, W. J. 2006. "Control of Lp(a) with diet and supplements." *Atherosclerosis* 7 (Suppl. 3):530.

5. Galloway, J. 2002. *Galloway's Book on Running,* 2nd ed. Bolinas, Calif.: Shelter Publications.

6. Schott, J., McCully, K., et al. 1995. "The role of metabolites in strength training." *European Journal of Applied Physiology* 71:337–41.

7. Cummings, D. E., Weigle, D. S., et al. 2002. "Plasma ghrelin levels after diet-induced weight loss or gastric bypass surgery." *New England Journal of Medicine* 346:1623–30.

8. Knott, C. D. 1998. "Changes in orangutan caloric intake, energy balance, and ketones in response to fluctuating fruit availability." *International Journal of Primatology* 19:1061–79.

9. Spiegel, K., Tasali, E., et al. 2004. "Sleep curtailment in healthy young men is associated with decreased leptin levels, elevated ghrelin levels, and increased hunger and appetite." *Annals of Internal Medicine* 141:846–50.

10. Andrews, M. T. 2007. "Advances in molecular biology of hibernation in mammals." *Bioessays* 29:431–40.

11. Buxton, O. 2007. "Body rhythms and mental states." In C. Wade; *Psychology,* 9th ed, San Francisco: Prentice Hall.

12. Ferraz, M. R., Ferraz, M. M., et al. 2001. "How REM sleep deprivation and amantadine affects male rat sexual behavior." *Pharmacology of Biochemical Behaviour* 3–4:325–32.

Chapter 9: BEGIN THE RESTORATION

1. Gundry, S. R., and Ehrman, W. J. 2006. "Diet and supplement effects on lipid subfractions and inflammatory markers in the metabolic syndrome: Change is possible." *Atherosclerosis* 7 (Suppl. 3):440.

2. Campbell, T. C. 2004. *The China Study.* Dallas: BenBella Books.

3. Fraser, G. E. 1999. "Association between diet and cancer, ischemic heart disease, and all-cause mortality in non-Hispanic, white California Seventh-Day Adventists." *American Journal of Clinical Nutrition* 70:532S–38S.

4. Fraser, G. E., and Shavlik, D. J. 2001. "Ten years of life: Is it a matter of choice?" *Archives of Internal Medicine* 161(13):1645–52.

5. Andrews, M. T. 2007. "Advances in molecular biology of hibernation in mammals." *Bioessays* 29:431–40.

6. Weigle, D. S., and Cummings, D. E. 2003. "Roles of leptin and ghrelin in the loss of body weight caused by a low fat, high carbohydrate diet." *Journal of Clinical Endocrinology and Metabolism* 88(4):1577–86.

Chapter 10: PICKING UP THE PACE

1. Ross, R., Dagnone, D., et al. 2000. "Reduction in obesity and related co-morbid conditions after diet-induced weight loss or exercise-induced weight loss in men." *Annals of Internal Medicine* 133:92–103.

2. Janssen, I., Fortier, A., et al. 2002. "Effects of an energy-restrictive diet with or without exercise on abdominal fat, intermuscular fat, and metabolic risk factors in obese women." *Diabetes Care* 25(3):431–38.

3. Ross, R., Janssen, I., et al. 2004. "Exercise-induced reduction in obesity and insulin resistance in women: a randomized controlled trial." *Obesity Research* 12(5):789–98.

4. Kopelman, P. G. 2001. *The Management of Obesity and Related Disorders.* London: Informa Health Care.

5. Hodgson, J. H., Watts, G. F., et al. 2002. "Coenzyme Q10 improves blood pressure and glycaemic control: a controlled trial in subjects with type 2 diabetes." *European Journal of Clinical Nutrition* 56(11):1137–42.

6. Loster, H., Miehle, K., et al. 1999. "Prolonged oral L-carnitine substitution increases bicycle ergometer performance in patients with severe, ischemically induced cardiac insufficiency." *Cardiovascular Drugs and Therapy* 13(6):537–46.

Chapter 11: THRIVING FOR A GOOD, LONG TIME

1. Sonneborn, J. S. 2005. "The myth and reality of reversal of aging by hormesis." *Annals of the New York Academy of Science* 1057:165-76.

2. Calabrese, E. J., and Baldwin, E. J. 1999. "Radiation hormesis: Origins, history, scientific foundations." *BELLE* 8(2): December 1999. Also at www.belleonline.com.

3. Sonneborn, J. S., and Barbee, S. C. 1998. "Exercise-induced stress response as an adaptive tolerance strategy." *Environmental Health Perspectives* 106 (Suppl. 1):325-30.

4. Walford, R. 2000. *Beyond the 120 Year Diet.* New York: Four Walls Eight Windows.

5. Larson, B. T. 2005. "When does 'living a dog's life' become a good option for humans?" *Journal of Nutrition, Health, and Aging* 9:135.

6. Glade, M. J. 2001. "Benefits from caloric restriction: Is it hormesis?" *Nutrition* 17(1):78-82.

7. Neese, R. M., and Williams, G. C. 1994. *Why We Get Sick.* New York: Vintage.

8. Samaras, T. J., and Elrick, H. 2002. "Height, body size and longevity: Is smaller better for the human body?" *Western Medical Journal* 17:206-08.

Chapter 12: TRICKING YOUR GENES: BEYOND DIET

1. Sonneborn, J. S. 2005. "The myth and reality of reversal of aging by hormesis." *Annals of the New York Academy of Science* 1057:165-76.

2. Glade, M. J. 2001. "Benefits from caloric restriction: Is it hormesis?" *Nutrition* 17(1):78-82.

3. Civitarese, D. E., and Carling, S. 2007. "Calorie restriction increases muscle mitochondrial biogenesis in healthy humans." *PLOS Medicine* 4(3):e76.

4. McClure, B. S., May, H. T., et al. 2007. "Fasting, a novel indicator of religiousity, may reduce the risk of coronary artery disease." *Circulation* 116:II_826-II_827.

5. Allebeck, P., and Rydberg, U. 1998. "Risks and protective effects of alcohol on the individual." *Alcoholism* 22:269s-79s.

6. Makamal, K. J., Chiuve, S. E., et al. 2006. "Alcohol consumption and risk of coronary heart disease in men with healthy lifestyles." *Archives of Internal Medicine* 166:2145-50.

7. Strange, S., Wu, T., et al. 2004. "Relationship of alcohol drinking pattern to risk of hypertension." *Hypertension* 44:813.

8. Ranheim, T., and Halvorsen, B. 2005. "Coffee consumption and human health—beneficial or detrimental?" *Molecular and Nutritional Food Research* 49(3):274-84.

9. Wilmoth, J. R., Deegan, L. J., et al. 2000. "Increase of maximum life-span in Sweden, 1861-1999," *Science* 289(5488):2366-68.

10. Abudu, N., Miller, J. J., et al. 2004. "Vitamins in human arteriosclerosis with emphasis on Vitamin C and Vitamin E." *Clinica Chimica Acta* 339:11-25.

11. Vuth, R. 1990. "The mechanism of Vitamin D toxicity." *Bone and Mineral* 11(3):267-72.

12. Jenkins, M. Y., and Mitchell, G. V. 1975. "The influence of excess Vitamin E on Vitamin A toxicity in rats." *Journal of Nutrition* 105:1600-06.

13. Chwelatiuk, E., Wlostowsk, T., et al. 2005. "Melatonin increases tissue accumulation and toxicity of cadmium in the bank vole." *BioMetals* 18:283-91.

14. Conti, B., Sanchez-Alavez, M., et al. 2006. "Transgenic mice with a reduced core body temperature have an increased life span." *Science* 314:824-28.

ACKNOWLEDGMENTS

This book, which traces my transformation from an obese heart surgeon to a practitioner of restorative medicine—as well as the resulting impact it has already had on thousands of individuals—would not have happened without an almost chance encounter. I was still Professor and Chairman of Cardiothoracic Surgery at Loma Linda University Medical Center when I met "Big Ed." Research had long governed my professional life, so the dramatic and positive changes in Big Ed's coronary angiogram after six months of weight loss and his seemingly random choice of nutriceutical supplements caused me to literally look again (the definition of research). Thank you, Ed. This reexamination included not only everything I had heretofore believed about disease causation, but also my Yale University undergraduate thesis; titled "Human Biological and Social Evolution," which was written under the direction of my first mentor, David Pilbeam.

With the unswerving encouragement of my wife, Penny, I embarked on a series of experiments on myself that produced equally startling results. But without the help of my then longtime administrative assistant, Barbara Wickham, and my nurse practitioner, Nancy Pike, who both also volunteered as guinea pigs, I doubt that this book would have ever been written. It was Barbara who prevailed on me to write detailed instructions on Diet Evolution for future patients, the very same instructions we continue to use. The changes in Barbara and Nancy's weight and health were so dramatic that we all knew something special was happening. And a special thanks to my "right hand" Chris Alvarez, nurse extraordinaire, who not only lost 26 pounds herself, but daily keeps all The Club members in-line with her advice and support.

Thanks go also to the now thousands of patients, who soon became known as members of The Club. They allowed me to draw and analyze their blood profiles every three months and gave me feedback on what worked and what didn't in the Diet Evolution program.

The actual book that you hold in your hands would not be possible without the tireless efforts of Olivia Bell Buehl, who took a wordy tome and through her word-smithing magic, got the best out of me (and herself). It would have never gotten to that point without the help of my agent, Bret Saxon, of Transactional Marketing Partners, who introduced me to Executive Editor Heather Jackson. Her sense that "The Next Great Diet" was lurking in my original manuscript propelled her to have me focus my attentions on the universal question, and one I hear every day: "What should I eat?" This book does just that. Thank you, Heather.

Thanks also to the rest of the phenomenal team of professionals at Crown Publishers and Random IIouse who have transformed mere pieces of paper to the exciting book you have in your hands!

My encounter with Bret would never have happened without my very good friend and patient Earl Greenburg passing my manuscript on to him with the warning to me that Bret hated every submission he had ever read. An hour later, Bret proved Earl wrong! The rest is history.

Thanks also to my partner, Dr. Walter Ehrman, former Professor of Surgery at Loma Linda University, who followed me to Palm Springs to establish The International Heart and Lung Institute. Λ consummate thoracic surgeon, Walter has allowed me to travel the world presenting the results of my diet interventions, while he is available to tend to the health of our patients. The works of Dr. Richard Dawkins and Dr. Robert Sapolsky aided in the development of my theories of disease causation and our built-in genetic programming. Although I have never met either author, I hope that knowing smiles will come to their faces as they read this book.

I have had several special mentors in my career as a cardiothoracic surgeon, from Dr. Andrew "Glenn" Morrow at the National Institutes of Health to Dr. Marvin Kirsch and Dr. Richard Burney at the University of Michigan, Dr. Joseph McLaughlin at the University of Maryland, Mr. Marc DeLeval at The Hospital for Sick Children in London, and my longtime partner and friend Dr. Leonard Bailey of Loma Linda University Medical Center. My most heartfelt thanks go out to Dr. Robert and Dr. Lois Ellison, my teachers at the Medical College of Georgia. Not only did they steer my course through medical school and beyond, their steadfast interest, genuine concern, and delight in my thirty-year career since graduation have been a source of unceasing encouragement. I hope this "paper" meets with their approval, as well. Thank you one and all.

RECIPE INDEX

ONE LAST WORD ...

At the start of this book, I shared some stories of others like you who desired a new way of eating and living, but didn't know what to do or what to expect from Diet Evolution. Here's where they are today.

Burt Kaplan

It's two years now that Burt has been practicing Diet Evolution. Nearing his 79th birthday, most of his old friends walk past him, mistaking him for a man 20 years younger. When they learn his identity, they also become my patients. His diabetes is gone and the medications are disappearing; Burt gets younger every year.

Before Diet Evolution

Weight: 226 pounds
Body Mass Index (BMI): 34
Body Fat Percentage: 29 percent
Total Cholesterol: 178 (on statins)
C-Reactive Protein: 8.8

After Diet Evolution

Weight: 175 pounds
Body Mass Index: 26
Body Fat Percentage: 19 percent
Total Cholesterol: 149
C-Reactive Protein: 1.1

Judith Rhode

One year later, Judith strides into my office with a happy grin. Down 30 pounds, she's nearly off insulin shots, active, walking without a walker, and bragging about how her children and husband have collectively lost over 110 pounds, thanks to Diet Evolution. (See Judith's stats on the following page.)

Before Diet Evolution

Weight: 163 pounds
Body Mass Index: 32
Body Fat Percentage: 42 percent
Blood Pressure: 150/70
Total Cholesterol: 239
C-Reactive Protein: 9.6

After Diet Evolution

Weight: 133 pounds
Body Mass Index: 26
Body Fat Percentage: 37 percent
Blood Pressure: 110/60
Total Cholesterol: 153
C-Reactive Protein: 1.4

Sandra Hall

I met Sandra for the first time a year after she had started Diet Evolution. Down 57 pounds, with normal blood pressure, she brought me her "before" photo. She just wanted to say thanks and to let me know that Diet Evolution worked. She's already learned that Diet Evolution is a lifestyle she can live with, without drugs.

Note: I don't have all her stats since she has never seen me as a patient (she doesn't need to now; she's got Diet Evolution!).

Before Diet Evolution

Weight: 249 pounds
Body Mass Index: 43
Total Cholesterol: 212

After Diet Evolution

Weight: 192 pounds
Body Mass Index: 32
Total Cholesterol: 134

Margo Hamilton

Margo dropped a pound every other day for the first eight weeks—all without exercise. She had no cravings, was shocked that she wasn't hungry, and often forgot to eat! The guidelines made restaurant eating a cinch; as promised, after ninety days, Margo became a different person . . .

Before Diet Evolution

Weight: 255 pounds
Body Mass Index: 41
Body Fat Percentage: 47 percent
Blood Pressure: 150/90
Total Cholesterol: 276
C-Reactive Protein: 4.2

After Diet Evolution

Weight: 213 pounds
Body Mass Index: 36
Body Fat Percentage: 43 percent
Blood Pressure: 140/80
Total Cholesterol: 226
C-Reactive Protein: 1.2

INDEX

ABOUT THE AUTHOR

Steven R. Gundry, M.D., is a cum laude graduate of Yale University with special honors in Human Biological and Social Evolution. After graduating Alpha Omega Alpha from the Medical College of Georgia School of Medicine, Dr. Gundry completed residencies in General Surgery and Thoracic Surgery at the University of Michigan and served as a Clinical Associate at the National Institutes of Health. There, he invented devices that reverse the cell death seen in acute heart attacks; variations of these devices subsequently became the Gundry Retrograde Cardioplegia Cannula. It has become the world's most widely used device of its kind to protect the heart from damage during open-heart surgery. After completing a fellowship in congenital heart surgery at The Hospital for Sick Children, Great Ormond Street, in London, and two years as a professor at The University of Maryland School of Medicine, Dr. Gundry was recruited as Professor and Chairman of Cardiothoracic Surgery at Loma Linda University Medical Center. There, he and his partner, Dr. Leonard Bailey, pioneered infant and pediatric heart transplantation. Together, they have performed more such transplants than any other surgeons in the world.

During his tenure at Loma Linda, Dr. Gundry pioneered the field of xenotransplantation, the study of how the genes of one species react to the transplanted heart of a foreign species. He was one of the original twenty investigators of the first FDA-approved implantable left ventricular assist device (a kind of artificial heart). Dr. Gundry is also the inventor of the Gundry Ministernotomy, the most widely used minimally invasive techique for operating on the aortic or mitral valves; the Gundry Lateral Tunnel, a "living" tissue that can rebuild parts of the heart in children with severe congenital heart malformations; and the Skoosh venous cannula, the most widely used cannula in minimally invasive heart operations.

One of the fathers of robotic surgery, as a consultant to Computer Motion (now Intuitive Surgical), Dr. Gundry received early FDA approval to use robotic-assisted minimally invasive surgery for coronary artery-bypass and mitral-valve operations. He holds patents on connecting blood vessels and coronary artery bypasses without sutures, as well as on repairing the mitral valve without the need for sutures or a heart-lung machine. He has served on the Board of Directors of the American Society of Artificial Internal Organs (ASAIO) and was a founding board member and treasurer of the International Society of Minimally Invasive Cardiothoracic Surgery (ISMICS). He recently completed two successive elected terms as President of the Board of Directors of the American Heart Association, Desert Division.

Dr. Gundry has been elected a Fellow of the American College of Surgeons, the American College of Cardiology, the American Surgical Association, the American Academy of Pediatrics, and the College of Chest Physicians. He is a member of numerous other surgical and medical societies. He is also the author of more than three hundred articles, chapters, and abstracts in peer-reviewed journals on surgical, immunologic, genetic, nutritional, and lipid investigations. He has operated in more than thirty countries, including charitable missions to China, India, and Zimbabwe.

Inspired by the stunning reversal of coronary artery disease in an "inoperable" patient, using a combination of dietary changes and nutriceutical supplements, in 2001, Dr. Gundry changed the path of his career. An obese, chronic "diet" failure himself, he adapted his undergraduate Yale University thesis to design a diet based on evolutionary genetic coding, which enabled him to reverse his own medical problems. In the process, he effortlessly lost 75 pounds. The equally astonishing results from following what he came to call Diet Evolution in several of his staff led Dr. Gundry to accept a position in Palm Springs where he could devote his efforts to disease reversal.

No longer satisfied with repairing the damage of chronic diseases, since 2002, Dr. Gundry has served as Medical Director of The International Heart and Lung Institute in Palm Springs, California, which serves patients referred from across the nation. In addition, he also founded and serves as Director of The Center for Restorative Medicine, which is part of the Institute. Its mission is to prevent and reverse the chronic diseases of "aging" with diet and nutriceutical interventions, using surgical intervention for heart and vascular disease as a last resort.

Dr. Gundry lives with his wife, Penny, and their four dogs, George, Bella, Chessie, and Fanny Foo Foo, in Palm Springs and Montecito, California. His two adult daughters live nearby.